T0325822

Taxonomic Guide to Infectious Diseases

Understanding the Biologic Classes of Pathogenic Organisms

For Jessie and Ellen

Taxonomic Guide to Infectious Diseases

Understanding the Biologic Classes of Pathogenic Organisms

Jules J. Berman

AMSTERDAM • BOSTON • HEIDELBERG • LONDON
NEW YORK • OXFORD • PARIS • SAN DIEGO
SAN FRANCISCO • SINGAPORE • SYDNEY • TOKYO
Academic Press is an imprint of Elsevier

Academic Press is an imprint of Elsevier
32 Jamestown Road, London NW1 7BY, UK
225 Wyman Street, Waltham, MA 02451, USA
525 B Street, Suite 1800, San Diego, CA 92101-4495, USA

First edition 2012

British Library Cataloguing-in-Publication Data
A catalogue record for this book is available from the British Library

Library of Congress Cataloging-in-Publication Data
A catalog record for this book is available from the Library of Congress

ISBN: 978-0-12-415895-5

For information on all Academic Press publications
visit our website at elsevierdirect.com

Typeset by MPS Limited, Chennai, India
www.adi-mps.com

Printed and bound in United States of America

12 13 14 15 16   10 9 8 7 6 5 4 3 2 1

Contents

Appendices

Preface

"Order and simplification are the first steps toward the mastery of a subject."

Thomas Mann

This book explains the biological properties of infectious organisms in terms of the properties they inherit from their ancestral classes. For example, the class of organisms known as Apicomplexa (Chapter 19) contains the organisms responsible for malaria, babesiosis, cryptosporidiosis, cyclosporan gastroenteritis, isosporiasis, sarcocystosis, and toxoplasmosis. When you learn the class properties of the apicomplexans, you'll gain a basic understanding of the biological features that characterize every infectious organism in the class (See Glossary item, Class).

If you are a student of microbiology, or a healthcare professional, you need to be familiar with hundreds of infectious organisms. There are many resources, web-based and paper-based, that describe all of these diseases in great detail, but how can you be expected to integrate volumes of information when you are confronted by a sick patient? It is not humanly possible. A much better strategy is to learn the basic biology of the 40 classes of organisms that account for all of the infectious diseases that occur in humans. After reading this book, you will be able to fit newly acquired facts, pertaining to individual infectious species, onto an intellectual scaffold that provides a simple way of understanding their clinically relevant properties.

Biological taxonomy is the scientific field dealing with the classification of living organisms. Non-biologists who give any thought to taxonomy, may think that the field is the dullest of the sciences. To the uninitiated, there is little difference between the life of a taxonomist and the life of a stamp collector. Nothing could be further from the truth. Taxonomy has become the grand unifying theory of the biological sciences. Efforts to sequence the genomes of prokaryotic, eukaryotic, and viral species, thereby comparing the genomes of different classes of organisms, have revitalized the field of evolutionary taxonomy (phylogenetics). The analysis of normal and abnormal homologous genes in related classes of organisms have inspired new disease treatments targeted against specific molecules and pathways characteristic of species or classes or organisms. Students who do not understand the principles of modern taxonomy have little chance of perceiving the connections between medicine, genetics, pharmacology, and pathology, to say nothing of clinical microbiology.

Here are some of the specific advantages of learning the taxonomy of infectious diseases.

1. AS A METHOD TO DRIVE DOWN THE COMPLEXITY OF MEDICAL MICROBIOLOGY

Learning all the infectious diseases of humans is an impossible task. As the number of chronically ill and immune-compromised patients has increased, so has the number of opportunistic pathogens. As global transportation has become commonplace, the number of exotic infections spread worldwide has also increased (see Glossary item, Exotic diseases in the United States). A few decades ago, infectious disease experts were expected to learn a few hundred infectious diseases. Today, there are over 1400 organisms that can cause diseases in humans, and the number is climbing rapidly, while the techniques to diagnose and treat these organisms are constantly improving. Textbooks cannot cover all these organisms in sufficient detail to provide healthcare workers with the expertise to provide adequate care to their patients.

How can any clinician learn all that is needed to provide competent care to patients? The first step in understanding infectious diseases is to understand the classification of pathogenic organisms. Every known disease-causing organisms has been assigned to one of 40 well-defined classes of organisms, and each class fits within a simple ancestral lineage. This means that every known pathogenic organism inherits certain properties from its ancestral classes and shares these properties with the other members of its own class. When you learn the class properties, along with some basic information about the infectious members of the classes, you gain a comprehensive understanding of medical microbiology.

2. TAXONOMY AS WEB COMPANION

"Getting information off the Internet is like taking a drink from a fire hydrant."

Mitchell Kapor

The web is a great resource. You can find a lot of facts, and if you encounter an unfamiliar word or a term, the web will provide a concise definition, in a jiffy. The web cannot, however, provide an understanding of the related concepts that form the framework of a scientific discipline. The web supplies facts, but books tell you what the facts mean.

Before the web, scientific texts needed to contain narrative material as well as the detailed, raw information pertaining to the field. For example, a microbiology text would be expected to contain long descriptions of each infectious organism, the laboratory procedures required to identify the organism, its clinical presentation and its treatment. As a result, authors were caught between writing enormous texts that contained much more information than any student could possibly absorb, or writing short works covering a narrow topic in microbiology, or writing review books that hinted at many different topics. Today, authors have the opportunity to create in-depth and comprehensive works that are quite short, without sacrificing conceptual clarity. The informational details can be deferred to the web! This book concentrates on its primary goal; describing all pathogenic organisms in relation to their taxonomic assignments. You will notice that for a relatively short text, the *Taxonomic Guide to Infectious Diseases* has a large index. The index was designed as a way to connect terms and concepts that appear in multiple places within the text, and as a key to information on the web. Most of the index terms have excellent coverage in Wikipedia. You will find that the material retrieved from Wikipedia will make much more sense to you, and will have much more relevance to your own professional activities, after you have read this book.

3. AS PROTECTION AGAINST PROFESSIONAL OBSOLESCENCE

There seems to be so much occurring in the biological sciences, it is just impossible to keep on top of things. With each passing day, you feel less in tune with modern science, and you wish you could return to a time when a few fundamental principles grounded your chosen discipline. You will be happy to learn that science is all about finding generalizations among data or among connected systems (i.e., reducing the complexity of data or finding simple explanations for systems of irreducible complexity). Much, if not all, of the perceived complexity of the biological sciences derives from the growing interconnectedness of once-separate disciplines: cell biology, ecology, evolution, climatology, molecular biology, pharmacology, genetics, computer sciences, paleontology, pathology, statistics, and so on. Scientists today must understand many different fields, and they must be willing and able to absorb additional disciplines, throughout their careers. As each field of science becomes entangled with others, the seemingly arcane field of biological taxonomy has gained in prominence because it occupies the intellectual core of virtually every biological field.

Modern biology seems to be data-driven. A deluge of organism-based genomic, proteomic, metabolomic and other "omic" data is flooding our databanks and drowning our scientists. These data will have limited scientific value if we cannot find a way to generalize the data collected for each organism to the data collected in other organisms. Taxonomy is the scientific method that reveals how different organisms are related. Without taxonomy, data has no biological meaning.

The discoveries that scientists make in the future will come from questions that arise during the construction and refinement of biological taxonomy. In the case of infectious diseases, when we find a trait that informs us that what we thought was a single species is actually two species, we can develop treatments optimized for each species. When we correctly group organisms within a common class, we can test and develop new drugs that are effective against all of the organisms within the class, particularly if those organisms are characterized by a molecule, pathway, or trait that is specifically targeted by a drug. Terms used in diverse sciences, such as homology, metabolic pathway, target molecule, acquired resistance, developmental stage, cladistics, monophyly, model organism, class property, phylogeny, all derive their meaning and their utility from biological taxonomy (see Glossary items, Cladistics, Monophyletic class). When you grasp the general organization of living organisms, you will understand how different scientific fields relate to each other, thus avoiding professional obsolescence.

HOW THE TEXT IS ORGANIZED

If you are reading *Taxonomic Guide to Infectious Diseases* to gain a general understanding of taxonomy, as it applies to human diseases, you may choose to read the introductory chapters, followed by reading the front sections of each subsequent chapter. You can defer reading the lists of infectious species until you need to relate general knowledge of a class of organisms to specific information on pathogenic species. If you are a healthcare professional, you will find that when you use the index to find the chapter that lists a particular organism or infectious disease, you will easily understand the disease because you have previously learned the important features of its biologic class. This deep knowledge will help you when you use other resources to collect detailed pathologic, clinical and pharmacologic information.

Though about 334 living organisms account for virtually all of the infectious diseases occurring in humans, about 1000 additional organisms account for "case report" incidents, involving one or several people, or an isolated geographic region, or otherwise-harmless organisms that cause disease under special circumstances. Appendix III lists nearly every known infectious organism (about 1400 species), and the taxonomic hierarchy for each genus. When you encounter the name of an organism, and you just can't remember anything about its taxonomic lineage (i.e., the class of the organism and the ancestral classes), you can find it quickly in the appendix. With this information, you can open the chapter that describes the class properties that apply to the species.

Some clinical concepts are taxonomically promiscuous. For example, the hepatitis viruses (A through G), are dispersed under several different classes of viruses. Moreover, the A through G list of hepatitis viruses excludes some of the most important viruses that target the liver (e.g., yellow fever virus, Dengue virus, Epstein–Barr virus). Topics that cross class boundaries, such as hepatitis viruses, long branch attraction, virulence factors, vectors, zoonoses, and many others, are included in the Glossary.

Nota Bene

Biological nomenclature has changed a great deal in the past few decades. If you learned medical microbiology in the preceding millennium, you may be surprised to learn that kingdoms have fallen (the once mighty kingdom of the protozoans has been largely abandoned), phyla have moved from one kingdom to another (the microsporidians, formerly protozoans, are now fungi), and numerous species have changed their names (*Pneumocystis carinii* is now *Pneumocystis jiroveci*). Most striking is the expansion of the existing ranks. Formerly, it was sufficient to divide the classification into a neat handful of divisions: Kingdom, Phylum, Class, Order, Family, Genus, Species. Today, the list of divisions has nearly quadrupled. For example, Phylum has been split into the following divisions: Superphylum, Phylum, Subphylum, Infraphylum, and Microphylum. The other divisions are likewise split. The subdivisions often have a legitimate scientific purpose. Nonetheless, current taxonomic order is simply too detailed for readers to memorize. Taxonomists referring to a class of any rank will sometimes use the word "taxon". I find this term somewhat lacking because it cannot be modified to refer to a direct parent or child taxon. In this book, all ranks will simply be referred to as "Class". The direct father class is the superclass, and the direct child class is the subclass. The terms "genus" (plural "genera") and "species" will preserve the binomial assignment of organism names. In the case of viruses, Baltimore Classification is used, which places every virus into one of seven Groups. Since "Group" is applied universally and consistently by virologists who employ the Baltimore Classification, its use is preserved here. Subdivisions of the Baltimore Group viruses are referred to herein as classes.

The use of "Class", "Superclass", and "Subclass" conforms to nomenclature standards developed by the metadata community (i.e. uses a standard terminology employed by the computational field dealing with the description of data) [1]. This simplified terminology avoids the complexities endured by traditional taxonomists. Regarding the use of upper and lower case terminology, when referring to a formal taxonomic class, positioned within the hierarchy, the uppercase letters and Latin plural forms are used (e.g. Class Eukaryota). When referring to the noun and adjectival forms, lowercase characters and the English pluralized form are used (e.g., a eukaryote, the eukaryotes, or eukaryotic organisms).

Each chapter contains a hierarchical listing of organisms, roughly indicating the ordered rank of the infectious genera covered in each chapter.

Classes that do not contain infectious organisms are omitted from the schema. Traditionally, the class rank would be listed in the hierarchy (e.g. Order, Suborder, Infraorder). In this book, the relative descent through the hierarchy is indicated by indentation. The lowest subclass in each taxonomic list is "genus", which is marked throughout with an asterisk. This visual method of ranking the classification produces an uncluttered, disease-only taxonomy and provides an approximate hierarchical rank for each class and species.

About the Author

Jules Berman received two bachelor degrees, in Mathematics and in Earth and Planetary Sciences, from MIT. He received his PhD from Temple University, and his MD from the University of Miami. He received post-doctoral training in the U.S. National Cancer Institute, and residency training at George Washington University Medical Center. He is board certified in Anatomic Pathology and in Cytopathology. He has served as Chief of Anatomic Pathology at the Baltimore Veterans Administration Medical Center. He was an Associate Professor at the University of Maryland School of Medicine, a Visiting Clinical Assistant Professor at the Johns Hopkins Medical Institutions, and a Medical Officer at the National Institutes of Health. He served as President of the Association for Pathology Informatics, in 2006, and received the Association for Pathology Informatics Lifetime Achievement Award, in 2011. Jules Berman is now a freelance writer and editor. He has first-authored seven science books, and over 100 journal articles.

About the Cover Art

The cover image, entitled "Taxonomic Instability" was composed by Jules Berman. The far-left sphere contains a micrographic image of *Escherichia coli*, photographed by Eric Erbe, and provided to the public by the U.S. Department of Agriculture. Moving clockwise, the small sphere in the far background is a scanning electron micrograph of intestinal mucosa, upon which a single, flattened Giardia trophozoite is tightly adhesed. The photograph was taken by Dr. Stan Erlandsen and provided to the public by the U.S. Centers for Disease Control and Prevention. Continuing clockwise, the upper right-side sphere is a color-enhanced scanning electron micrograph of *Salmonella typhimurium* (purple) invading cultured cells. The photograph was taken at the Rocky Mountain Laboratories and has been provided to the public by the U.S. National Institute for Allergy and Infectious Diseases. The sphere on the far right, contains drawings of fungi of Class Basidiomycota, taken from Plate 63 of Ernst Haeckel's book *Kunstformen der Natur*, published in 1904 (now in the public domain). The sphere in the low foreground contains a scanning electron micrograph of *Staphylococcus aureus* with intermediate level resistance to vancomycin. The photograph was taken by Janice Haney Carr and provided to the public by the U.S. Centers for Disease Control and Prevention. The aggregate scene was composed in a script written by Jules Berman, in the Pov-Ray Scene Description Language. General techniques for describing cloudscapes, for Pov-Ray, were explained in a tutorial by Friedrich A. Lohmuellerr, at: www.f-lohmueller.de. The script was rendered into the final image using the freely available Pov-Ray (tm), Persistence of Vision Raytracer Pty. Ltd., version 3.62, available at: www.povray.org.

Part I

Principles of Taxonomy

Chapter 1

The Magnitude and Diversity of Infectious Diseases

"All interest in disease and death is only another expression of interest in life."

Thomas Mann

THE IMPORTANCE OF INFECTIOUS DISEASES IN TERMS OF HUMAN MORTALITY

According to the U.S. Census Bureau, on July 20, 2011, the USA population was 311 806 379, and the world population was 6 950 195 831 [2]. The U.S. Central Intelligence agency estimates that the USA crude death rate is 8.36 per 1000 and the world crude death rate is 8.12 per 1000 [3]. This translates to 2.6 million people dying in 2011 in the USA, and 56.4 million people dying worldwide. These numbers, calculated from authoritative sources, correlate surprisingly well with the widely used rule of thumb that 1% of the human population dies each year.

How many of the world's 56.4 million deaths can be attributed to infectious diseases? According to World Health Organization, in 1996, when the global death toll was 52 million, "Infectious diseases remain the world's leading cause of death, accounting for at least 17 million (about 33%) of the 52 million people who die each year" [4]. Of course, only a small fraction of infections result in death, and it is impossible to determine the total incidence of infectious diseases that occur each year, for all organisms combined. Still, it is useful to consider some of the damage inflicted by just a few of the organisms that infect humans.

Malaria infects 500 million people. About 2 million people die each year from malaria [4].

About 2 billion people have been infected with *Mycobacterium tuberculosis*. Tuberculosis kills about 3 million people each year [4].

Each year, about 4 million children die from lung infections, and about 3 million children die from infectious diarrheal diseases [4]. Rotaviruses are one of many causes of diarrheal disease (Group III Viruses, Chapter 41).

J.J. Berman: Taxonomic Guide to Infectious Diseases. DOI: http://dx.doi.org/10.1016/B978-0-12-415895-5.00001-5

In 2004, rotaviruses were responsible for about half a million deaths, mostly in developing countries [5].

Worldwide, about 350 million people are chronic carriers of hepatitis B, and about 100 million people are chronic carriers of hepatitis C. In aggregate, about one quarter (25 million) of the hepatitis C chronic carriers will eventually die from ensuing liver diseases [4].

Infectious organisms can kill individuals through mechanisms other than through the direct pathologic effects of growth, invasion, and inflammation. Infectious organisms have been implicated in vascular disease. The organisms implicated in coronary artery disease and stroke include *Chlamydia pneumoniae* and Cytomegalovirus [6].

Infections caused by a wide variety of infectious organisms can result in cancer. About 7.2 million deaths occur each year from cancer, worldwide. About one-fifth of these cancer deaths are caused by infectious organisms [7]. In Europe, 60–70% of liver cancer cases are caused by hepatitis C virus; 10–15% of liver cancer is caused by hepatitis B infection [8]. Organisms contributing to cancer deaths include bacteria (*Helicobacter pylori*), animal parasites (schistosomes and liver flukes), and viruses (Herpesviruses, Papillomaviruses, Hepadnaviruses, Flaviviruses, Retroviruses, Polyomaviruses). Though fungal and plant organisms do not seem to cause cancer through human infection, they produce a multitude of biologically active secondary metabolites (i.e., synthesized molecules that are not directly involved in the growth of the organism), some of which are potent carcinogens. For example, aflatoxin produced by *Aspergillus flavus*, is possibly the most powerful carcinogen ever studied [9].

In summary, infectious diseases are the number one killer of humans worldwide, and they contribute to vascular disease and cancer, the two leading causes of death in the most developed countries. These observations clearly indicate that every healthcare professional, not just infectious disease specialists, must understand the biology of infectious organisms. A listing of the number of occurrences of some common infectious diseases is provided in Appendix II.

ONLY A SMALL PERCENTAGE OF TERRESTRIAL ORGANISMS ARE PATHOGENIC IN HUMANS

Given all the suffering caused by infectious organisms, you might begin to wonder whether the majority of terrestrial life-forms are devoted to the annihilation of the human species. Not to worry. Only a tiny fraction of the life forms on earth are infectious to humans. The exact fraction is hard to estimate because nobody knows the total number of terrestrial species. Most taxonomists agree that the number is in the millions, but estimates range from a few million up to several hundred million.

It is worth noting that species counts, even among the most closely scrutinized classes of organisms, are prone to underestimation. In the past, the rational basis for splitting a group of organisms into differently named species required, at the very least, heritable functional or morphologic differences among the members of the group. Gene sequencing has changed the rules for assigning new species. For example, various organisms with subtle differences from *Bacteroides fragilis* have been elevated to the level of species based on DNA homology studies. These include *Bacteroides distasonis*, *Bacteroides ovatus*, *Bacteroides thetaiotaomicron*, and *Bacteroides vulgatus* [10].

For the sake of discussion, let us accept that there are 50 million species of organisms on earth (a gross underestimate by some accounts). There have been about 1400 pathogenic organisms reported in the medical literature. This means that if you should stumble randomly upon a member of one of the species of life on earth, the probability that it is an infectious pathogen is about 0.000028.

Of the approximately 1400 infectious organisms that have been recorded somewhere in the medical literature, the vast majority of these are "case report" items; instances of diseases that have, to the best of anyone's knowledge, occurred once or a handful of times. They are important to epidemiologists because today's object of medical curiosity may emerge as tomorrow's global epidemic. Very few of these ultra-rare causes of human disease ever gain entry to a clinical microbiology textbook. Textbooks, even the most comprehensive, cover about three hundred organisms (excluding viruses) that are considered clinically important. In this book, we will cover about 350 living organisms and about 150 viruses within the main text. The Appendix lists about 1400 organisms (common, rare, and ultra-rare). This may seem like way too much to learn, but do not despair. Infectious agents fall into a scant 40 biological classes (32 classes of living organisms plus seven classes of viruses plus one class of prions). When you've learned the basic biology of the major taxonomic divisions that contain all the infectious organisms, you will understand the fundamental biological features that characterize every clinical organism. Almost everything else you need to learn can be acquired from web resources.

Chapter 2

What is a Classification?

"Deus creavit, Linnaeus disposuit," Latin for "God Creates, Linnaeus organizes."
Carolus Linnaeus

CLASSIFICATIONS DRIVE DOWN THE COMPLEXITY OF KNOWLEDGE DOMAINS

The human brain is constantly processing visual and other sensory information collected from the environment. When we walk down the street, we see images of concrete and asphalt and grass and other persons and birds and so on. Every step we take conveys a new world of sensory input. How can we process it all? The mathematician and philosopher Karl Pearson (1857–1936) has likened the human mind to a "sorting machine" [11]. We take a stream of sensory information and sort it into objects, and then we collect the individual objects into general classes. The green stuff on the ground is classified as "grass", and the grass is subclassified under some larger groups such as "plants". Flat stretches of asphalt and concrete may be classified under "road" and the road might be subclassified under "man-made constructions". If we did not have a culturally determined classification of objects in the world, we would have no languages, no ability to communicate ideas, no way to remember what we see, and no way to draw general inferences about anything at all. Simply put, without classification, we would not be human.

Every culture has some particular way to impose a uniform way of perceiving the environment. In English-speaking cultures, the term "hat" denotes a universally recognized object. Hats may be composed of many different types of materials, and they may vary greatly in size, weight, and shape. Nonetheless, we can almost always identify a hat when we see one, and we can distinguish a hat from all other types of objects. An object is not classified as a hat simply because it shares a few structural similarities with other hats. A hat is classified as a hat because it has a relationship to every other hat, as an item of clothing that fits over the head. Likewise, all biological classifications are built by relationships, not by similarities [12].

J.J. Berman: Taxonomic Guide to Infectious Diseases. DOI: http://dx.doi.org/10.1016/B978-0-12-415895-5.00002-7

GENERAL PRINCIPLES OF CLASSIFICATION

Oddly enough, despite the importance of classification in our lives, few humans have a rational understanding of the process of classification; it's all done for us on a subconscious level. Consequently, when we need to build and explain a formal classification, it can be difficult to know where to begin. As an example, how might we go about creating a classification of toys? Would we arrange the toys by color (red toys, blue toys, etc.), or by size (big toys, medium-sized toys), or composition (metal toys, plastic toys, cotton toys). How could we be certain that when other people create a classification for toys, their classification will be equivalent to ours?

For modern biologists, the key to the classification of living organisms is evolutionary descent (i.e., phylogeny). The hierarchy of classes corresponds to the succession of organisms that evolved from the earliest living organism to the current set of extant species. Historically, pre-Darwinian biologists who knew nothing about evolution, somehow produced a classification that looked much like the classification we use today. Before the discovery of the Burgess shale (discovered in 1909 by Charles Walcott), taxonomists could not conduct systematic reviews of organisms in rock strata; hence, they could not determine the epoch in which classes of organisms first came into existence, nor could they determine which fossil species preceded other species. Until late in the twentieth century, taxonomists could not sequence nucleic acids; hence, they could not follow the divergence of shared genes in different organisms. Yet they managed to produce a fairly accurate taxonomy. A nineteenth-century taxonomist would have no trouble adjusting to the classification used in this book.

How did the early taxonomists arrive so close to our modern taxonomy, without the benefit of the principles of evolution, geobiology, modern paleontological discoveries, or molecular biology? For example, how was it possible for Aristotle to know, about two thousand years ago, that a dolphin is a mammal, not a fish? Aristotle studied the anatomy and the developmental biology of many different types of animals. One large group of animals was distinguished by a gestational period in which a developing embryo is nourished by a placenta, and the offspring are delivered into the world as formed, but small versions of the adult animals (i.e., not as eggs or larvae), and the newborn animals feed from milk excreted from nipples, overlying specialized glandular organs (mammae). Aristotle knew that these were features that specifically characterized one group of animals and distinguished this group from all the other groups of animals. He also knew that dolphins had all these features; fish did not. He correctly reasoned that dolphins were a type of mammal, not a type of fish. Aristotle was ridiculed by his contemporaries for whom it was obvious that dolphins were a type of fish. Unlike Aristotle, they based their classification on similarities, not on relationships. They saw that dolphins looked like fish and dolphins swam in the ocean like fish, and this was all the proof they needed. For about two thousand years following the death of Aristotle, biologists persisted in their belief that dolphins were a type of fish. For the past several hundred years, biologists have acknowledged that Aristotle was correct after all; dolphins are mammals.

Aristotle, and legions of taxonomists that followed him, understood that taxonomy is all about finding the key properties that characterize entire classes and subclasses of organisms. Selecting the defining properties from a large number of morphologic, developmental and physiologic features in many different species requires attention to detail, and occasional moments of intellectual brilliance. To build a classification, the taxonomist must perform the following: (1) define classes (i.e., find the properties that define a class and extend to the subclasses of the class); (2) assign species to classes; (3) position classes within the hierarchy; and (4) test and validate all the above. These tasks require enormous patience and humility.

A classification is a hierarchy of objects that conforms to the following principles:

1. The classes (groups with members) of the hierarchy have a set of properties or rules that extend to every member of the class and to all of the subclasses of the class, to the exclusion of all other [unrelated] classes. A subclass is itself a type of class wherein the members have the defining class properties of the parent class plus some additional property(ies) specific for the subclass.
2. In a hierarchical classification, each subclass may have no more than one parent class. The root (top) class has no parent class. The biological classification of living organisms is a hierarchical classification.
3. At the bottom of the hierarchy is the species. In the classification of living organisms, the species is the collection of all the organisms of the same type (e.g., every squirrel belongs to a species of "squirrel").
4. Classes and species are intransitive. As examples, a horse never becomes a sheep, and Class Bikonta never transforms into Class Unikonta.
5. The members of classes may be highly similar to each other, but their similarities result from their membership in the same class (i.e., conforming to class properties), and not the other way around (i.e., similarity alone cannot define class inclusion).

It is important to distinguish a classification system from an identification system. An identification system matches an individual organism with its assigned object name (or species name, in the case of the classification of living organisms). Identification is based on finding several features that, taken together, can help determine the name of an organism. For example, if you have a list of identifiers: large, hairy, strong, African, jungle-dwelling, knuckle-walking; you might correctly identify the organism as a gorilla. These identifiers are different from the phylogenetic features that were used to classify gorillas within the hierarchy of organisms (Animalia: Chordata: Mammalia: Primates: Hominidae: Homininae: Gorillini: Gorilla). Specifically, you can identify an animal as a gorilla without knowing that a gorilla is a type of mammal. You can classify a gorilla as a member of Class Gorillini without knowing that a gorilla happens to be large. One of the most common mistakes in biology is to confuse an identification system with a classification system. The former provides a handy way to associate an object with a name; the latter is a system of relationships among organisms.

Still, it is impossible to forget that every species is populated by members that are similar to one another. When you see two squirrels playing in a tree, you cannot help but notice that the two squirrels are virtually identical. You can see their similarities, even if you cannot recite the biological features that define the squirrel species. You recognize them as squirrels because of their similarities to each other. You correctly infer that the two squirrels are characterized by a shared collection of "squirrel" genes. Likewise, the species of tiger is characterized by a set of genes that produce tigers. Every species on earth can be characterized by their genes. You go one step further, and you infer that if we had all the genes sequenced for every organism on the planet, we could create a complete and accurate classification for all the terrestrial life forms. Is this last inference correct? We will be discussing this topic in Chapter 3, The Tree of Life.

Chapter 3

The Tree of Life

"Individuals do not belong in the same taxon because they are similar, but they are similar because they belong to the same taxon."

George Gaylord Simpson (1902–1984) [13]

Bacteria (Chapters 1–14)
Eukaryota (Chapter 15)
  Bikonta (2-flagellum)
    Excavata (Chapters 16–18)
    Archaeplastida (Chapter 24)
    Chromalveolata (Chapters 19–21)
  Unikonta (1-flagellum)
    Amoebozoa (Chapter 22)
    Opisthokonta
      Choanozoa (Chapter 23)
      Animalia (Chapters 25–32)
      Fungi (Chapters 33–37)

Taxonomy is the science of classifying the elements of a knowledge domain, and assigning names to the classes and the elements. In the case of terrestrial life forms, taxonomy involves assigning a name and a class to every species of life. Biologists estimate that there are about 50 million living or extinct species, so the task of building a biological taxonomy is likely to continue for as long as humans dwell on earth. Most biologists would agree to the following:

1. All living organisms on earth contain DNA, a highly stable nucleic acid. DNA is transcribed into a less-stable, single-stranded molecule called RNA, and RNA is translated into proteins. All living organisms replicate their DNA and produce more organisms of the same genotype.
2. All living organisms on earth can be divided into two broad classes: the prokaryotes (organisms with a simple string of DNA and without a membrane-delimited nucleus, or any other membrane-delimited organelles),

J.J. Berman: Taxonomic Guide to Infectious Diseases. DOI: http://dx.doi.org/10.1016/B978-0-12-415895-5.00003-9

the class that includes all bacteria; and eukaryotes (organisms with a membrane-delimited nucleus) [14].

3. The prokaryotes preceded the emergence of the eukaryotes, and the first eukaryotes were built from the union of two or more prokaryotes [15].
4. Every eukaryotic organism that lives today is a descendant of a single eukaryotic ancestor [15].
5. Every organism belongs to a species that has a set of features that characterizes every member of the species and that distinguishes the members of the species from organisms belonging to any other species.

Of course, it is difficult to garner unanimous agreement by scientists, and every fundamental principle of taxonomy has been challenged at one time or another. For those who would include viruses (Chapters 38–45) and prions (Chapter 46) among the living organisms, statements 1 and 2 are debatable. The validity of statements 3 and 4 has been questioned. Some scientists have postulated that prokaryotes descended from eukaryotes, shucking off organelles in favor of a more simple, casual life-style. Others suggest that prokaryotes and eukaryotes arose simultaneously. In this case, any genetic or metabolic homologies between eukaryotes and prokaryotes result from a shared gene pool.

Statement 5 has a long and disputatious history. It has been argued that nature produces individuals, not species; the concept of species being a mere figment of the human imagination, created for the convenience of taxonomists who need to group similar organisms. There are those who would use computational methods to group organisms into various species. If you start with a set of feature data on a collection of organisms, you can write a computer program that will cluster the organisms into species, according to their similarities. In theory, one computer program, executing over a large dataset containing measurements for every earthly organism, could create a complete biological classification. The status of a species is thereby reduced from a fundamental biological entity to a mathematical construction.

This view is anathema to classic taxonomists, who have long held that a species is a natural unit of biological life, and that the nature of a species is revealed through the intellectual process of building a consistent taxonomy [16]. There are a host of problems consequent to computational methods for classification. First, there are many different mathematical algorithms that cluster objects by similarity. Depending on the chosen algorithm, the assignment of organisms to one species or another would change. Secondly, mathematical algorithms do not cope well with species convergence. Convergence occurs when two species independently acquire an identical or similar trait through adaptation; not through inheritance from a shared ancestor. Examples are: the wing of a bat and the wing of a bird; the opposable thumbs of opossums and of primates; the beak of a platypus and the beak of a bird. Unrelated species frequently converge upon similar morphologic solutions to common environmental conditions or shared physiological imperatives. Algorithms that cluster organisms based on similarity are likely to group divergent organisms under the same species.

It is often assumed that computational classification, based on morphologic feature similarities, will improve when we acquire whole-genome sequence data for many different species. Imagine an experiment wherein you take DNA samples from every organism you encounter: bacterial colonies cultured from a river, unicellular non-bacterial organisms found in a pond, small multicellular organisms found in soil, crawling creatures dwelling under rocks, and so on. You own a powerful sequencing machine, that produces the full-length sequence for each sampled organism, and you have a powerful computer that sorts and clusters every sequence. At the end, the computer prints out a huge graph, wherein all the samples are ordered, and groups with the greatest sequence similarities are clustered together. You may think you've created a useful classification, but you haven't really, because you don't know anything about the organisms that are clustered together. You don't know whether each cluster represents a species, or a class (a collection of related species), or whether a cluster may be contaminated by organisms that share some of the same gene sequences, but are phylogenetically unrelated (i.e., the sequence similarities result from chance or from convergence, but not by descent from a common ancestor). The sequences do not tell you very much about the biological properties of specific organisms, and you cannot infer which biological properties characterize the classes of clustered organisms. You have no certain knowledge whether the members of any given cluster of organisms can be characterized by any particular gene sequence (i.e., you do not know the characterizing gene sequences for classes of organisms). You do not know the genus or species names of the organisms included in the clusters, because you began your experiment without a presumptive taxonomy. Basically, you simply know what you knew before you started; that individual organisms have unique gene sequences that can be grouped by sequence similarity. A strictly molecular approach to classification has its limitations, but we shall see, in Chapter 4, that thoughtful biologists can use molecular data to draw profound conclusions about the classification of living organisms.

Taxonomists are constantly engaged in an intellectual battle over the principles of biological classification. They all know that the stakes are high. When unrelated organisms are mixed together in the same class, and when related organisms are separated into unrelated classes, the value of the classification is lost, perhaps forever. To understand why this is true, you need to understand that a classification is a hypotheses-generating machine. Species within a class tend to share genes, metabolic pathways, and structural anatomy. Shared properties allow scientists to form general hypotheses that may apply to all the members of a class. Without an accurate classification of living organisms, it would be impossible to make significant progress in the diagnosis, prevention, or treatment of infectious diseases.

James Joyce is credited with saying that "there are two sides to every argument; unfortunately, I can only occupy one of them." Students of the life sciences simply cannot hope to understand terrestrial organisms without accepting, at least tentatively, statements 1 through 5. After they have mastered the principles and practice of modern taxonomy, as described herein, they can re-assess the value of contrarian arguments.

Part II

Bacteria

Chapter 4

Overview of Class Bacteria

"The beginnings and endings of all human undertakings are untidy."

John Galsworthy

- Bacteria
  - Proteobacteria
    - Alpha Proteobacteria (Chapter 5)
    - Beta Proteobacteria (Chapter 6)
    - Gamma Proteobacteria (Chapter 7)
    - Epsilon Proteobacteria (Chapter 8)
  - Spirochaetes (Chapter 9)
  - Bacteroidetes (Chapter 10)
  - Fusobacteria (Chapter 10)
  - Firmicutes
    - Bacilli (Chapter 12)
    - Clostridia (Chapter 12)
    - Mollicutes (Chapter 11)
  - Chlamydiae (Chapter 13)
  - Actinobacteria (Chapter 14)

To understand the classification of bacteria, let us look at the controversy at the root of the tree of life. Three major groups of organisms account for all life on earth: eubacteria, archaea, and eukaryotes. When you compare species of Class Eubacteria with species of Class Archaea, you're not likely to notice any big differences. The eubacteria have the same shapes and sizes as the archaea. All species of eubacteria and all species of archaea are single-celled organisms, and they all have a typical prokaryotic structure (i.e., lacking a membrane-bound nucleus to compartmentalize their genetic material). As it happens, Class Eubacteria contains all of the bacterial organisms that are known to be pathogenic in humans. The Archaeans are non-pathogenic; most are extremophiles, capable of living in hostile environments (e.g., hot springs, salt lakes), but some Archaean species can occupy less-demanding biological niches (e.g., marshland, soil, human colon). Class Archaea does

J.J. Berman: Taxonomic Guide to Infectious Diseases. DOI: http://dx.doi.org/10.1016/B978-0-12-415895-5.00004-0

not hold a monopoly on extremophilic prokaryotes; some eubacterial species live in extreme environments (e.g. the alkaliphile *Bacillus halodurans*). The third major class of organisms, the Eukaryotes, is distinguished by the presence of a membrane-bound nucleus.

For many years, archaean species were considered just another class of bacteria. This changed in 1977. Woese and Fox had been studying ribosomal RNA. Because ribosomal RNA is a fundamental constituent of all living organisms, sequence comparisons in the genes coding for ribosomal RNA are considered a reliable way to estimate the degree of relatedness among organisms. In 1977, Woese and Fox surprised biologists when they demonstrated profound differences in the sequence of ribosomal RNA that distinguished archaean species from all other bacteria [17]. Much more shocking was their finding that the sequence of archaean ribosomal RNA was more closely related to eukaryotic cells than to other bacterial cells. Other sources have since shown that the archaeans share with the eukaryotes a variety of features that are lacking in the eubacteria. These include the presence of histones in some archaeans, the manner in which DNA is replicated and organized, and the finding that archaeans and eukaryotes share several transcription factors (proteins that bind to DNA and control the transcription of DNA to RNA) [18]. Woese and Fox proposed that the archaeans (then called archaebacteria) comprised a kingdom, separate but equal to the bacteria. That paper, and the many contributions of Woese that followed, sharpened our understanding of terrestrial biology and sparked a controversy that shook the foundations of taxonomic orthodoxy [14,19].

The solar system (sun, earth, moon, planets, and meteorites) all seem to have formed about 4.5 billion years ago. The first living organisms appeared about 3.8 billion years ago. Here are the hypothesized roots of terrestrial life, visualized through indentation:

Class Archaea
    Class Eubacteria
    Class Eukaryota

The first organisms to appear on earth, about 3.8 billion years ago, were Class Archaea, the root organism for the Tree of Life. From Class Archaea came Class Bacteria and Class Eukaryota. The argument for this particular schema is that many archaean species observed today are extremophiles that thrive under environmental conditions that are similar to the conditions on earth about 3.8 billion years ago. The members of Class Eubacteria are better-suited for a temperate, oxygen-rich atmosphere, such as we have today. Presumably, the eubacteria evolved after the archaeans appeared, from archaean ancestors. The eukaryotes, like all other living organisms on earth, replicate and transcribe DNA and translate RNA into proteins. Key structures of the genetic machinery of eukaryotes (chromosomal proteins and ribosomal RNA) are more similar to archaeans than to the eubacteria. Hence, it is presumed that the eukaryotes evolved from an archaean ancestor.

Here is another possible schema:

Class Archaea
  Class Eukaryota
Class Eubacteria

This schema is much like its predecessor (see above), but the eubacteria are not presumed to have developed from an archaean ancestor. Despite its name, archaea, from the Latin root meaning "ancient thing," there is no evidence that Class Archaea is older than Class Eubacteria. Profound differences between the DNA replication mechanisms in archaeans and eubacteria suggest that the two classes of bacteria may have arisen independently [15].

Alternately, Class Eubacteria and Class Archaea may have arisen from an ancient ancestor that has been lost to antiquity [15], as shown here:

Class Bacteria (eventually to contain Class Eubacteria, Class Archaea)
  Class Eukaryota

This schema was very popular prior to the work of Woese and colleagues, and still has many adherents. Here, we imagine that the first living organisms on earth were simple bacterial cells (cells without nuclei, also called prokaryotes or monera). As organisms evolved, they exchanged nuclear material somewhat promiscuously. Eventually two subclasses separated: Class Archaea and Class Eubacteria. It has been observed that most of the genes that code for basic cellular functions, such as enzymes, transport systems, cell wall synthesis, are quite similar in archaeans and in eubacteria [14]. This would suggest that the two classes of organisms shared a common ancestor. After the two classes of prokaryotes were established (maybe a billion years later), a third type of organism, characterized by a membrane-covered nucleus, was created through a felicitous merger of two or more cells, with contributions from Class Eubacteria and Class Archaea. This would explain why eukaryotic organisms have features of both Class Eubacteria and Class Archaea (discussed further in Chapter 15).

Alternately, as shown below, three classes of organisms may have arisen by a mechanism that is unknown and beyond the pale of credible scientific speculation.

Class Archaea
Class Eubacteria
Class Eukaryota

This agnostic schema gives Class Archaea, Class Eubacteria, and Class Eukaryota an equivalent stature at the root level of living organisms; thus removing precedence and ancestry for the top classes of life. There seems to be no doubt that all three major classes share certain genetic attributes, but no-one can say with certainty how these genetic attributes came to be shared or when each class arose. With this schema, there is a tacit acknowledgment that we do not know very much about the origins of life on earth.

In this book, we will not go into any of the arguments for or against any of the "origin of life" hypotheses. Readers can accept the observation that every serious evolutionary biologist is obsessed with the minutiae of these arguments; and rightly so. Biologists believe that the cell is a living book, recording the history and predicting the future of life on earth [20]. It is the responsibility of biologists to correctly interpret the contents of the book of life.

One of the most influential contributors to the field of molecular phylogenetics of bacteria is Carl Woese, who refined his earlier observations by using ribosomal RNA sequence variations as a biologic chronometer. By observing changes in the sequences of ribosomal RNA, he determined the branchings, over time, that occur during the evolution of organisms. Though he was looking at sequence similarities, his observations were focused on those similarities that revealed the sequential phylogenetic steps of bacterial evolution [21]. Furthermore, he built his classification on a pre-existing nomenclature that defined organisms on a species level.

Prior to the work of Woese and others, in the 1990s, there was little hope that the bacteria could be classified sensibly into a phylogenetic hierarchy. It was known that bacteria exchange genetic material horizontally, from one species of organism to another. During the several billion years of bacterial evolution, it was likely that primitive organisms merged with one another. There is every reason to believe that early viruses pulled and pushed fragments of DNA among many different bacterial organisms. It is even possible that native molecules of DNA, formed in the primordial soup of ancient earth, were copied and shared by members of different branches of evolving organisms. Biologists expected that these promiscuous exchanges of genetic material would have created a collection of different bacterial organisms for which a strict geneologic classification was impossible to construct. Indeed, generations of microbiologists tried and failed to produce a credible classification for bacteria. Because most bacteria fall into a narrow range of form and size, taxonomists lacked a sufficient set of morphologic features from which they could define and distinguish classes of organisms. They settled for a systematic grouping of bacteria based on a few morphologic features: dichotomizing staining properties (Gram positivity and Gram negativity), and a set of growth characteristics (e.g., nutrient requirements and the expression of traits expressed by colonies grown on various substrates).

It came as a great surprise to many when Woese and others developed a sensible classification of bacteria based on their analyses of a single molecule, rRNA. It remains to be seen whether Woese's one-molecule classification will withstand further scrutiny [22].

In the past decade, taxonomists have acquired access to the full genome sequences of many different organisms. The genome size of most bacteria fall in the range of 0.5 million base pairs up to about 10 million base pairs. This is a tiny fraction of the size of the human genome, which is about 3 billion base pairs in length. The organism with the largest genome is currently thought to be *Polychaos dubium* (Class Amoebozoa, Chapter 22), with a genome length of 670 billion base pairs. Because the bacterial genome is small, many of the first genomes to be fully sequenced belong to bacterial species, and dozens of full-length sequences are currently available to taxonomists.

It was hoped that comparisons between whole-genome sequences, on many different bacterial species, would solve many of the mysteries and controversies of bacterial taxonomy. These expectations were overly optimistic, due, in no small part, to an analytic phenomenon now known as "non-phylogenetic signal" [23]. When gene sequence data are analyzed, and two organisms share the same sequence in a stretch of DNA, it can be very tempting to infer that the two organisms belong to the same class (i.e., that they inherited the identical sequence from a common ancestor). This inference is not necessarily correct. Because DNA mutations arise stochastically over time (i.e., at random locations in the gene, and at random times), two organisms having different ancestors may achieve the same sequence in a chosen stretch of DNA. Conversely, if two organisms are closely related, there may be an identifiable ancestor, with the same DNA sequence found in one of the two organisms, that is not found in the other organism. When mathematical phylogeneticists began modeling inferences drawn from analyses of genomic data, they assumed that most class assignment errors would occur when the branches between sister taxa were long (i.e., when a long time elapsed between evolutionary divergences, allowing for many random substitutions in base pairs). They called this phenomenon, wherein non-sister taxa were assigned the same ancient ancestor class, "long branch attraction." In practice, errors of this type can occur whether the branches are long, or short, or in-between. Over the years, the accepted usage of the term "long branch attraction" has been extended to just about any error in phylogenetic grouping due to gene similarities acquired through any mechanism other than inheritance from a shared ancestor. This would include random mutations and adaptive convergences [24]. The moral here is that powerful data-intensive analytic techniques are sometimes more confusing than they are clarifying.

Though the field of computational taxonomy is flawed, readers must also understand that the field of classical taxonomy suffers from a self-referential paradox known as bootstrapping. Classical taxonomists need to have a classification of organisms before they can clearly see the relationships among classes (that is the purpose of a classification). Furthermore, taxonomists must see the relationships among classes before they can create the classification. Basically, a classification cannot be built without the assistance of a finished classification. In practical terms, taxonomists are continually constructing classes from their own scientific biases concerning the essential features of organisms. Classical taxonomists tend to be happy with class assignments that reinforce their original bias. Computational taxonomists insist that their approach overcomes the self-referential paradox by clustering similar organisms into classes based on an unbiased (though arbitrary) set of objectively measured features.

The wise taxonomist understands that new classifications are built upon old classifications, and that the value of any classification stems from our persistence in testing and revising tentative class assignments; not in our ability to re-compute a classification from a blank slate.

Here are a few suggestions on the best applications of molecular taxonomic techniques:

1. To provide some confirmation for a candidate class construction. You would expect key sequences among members of a class to be similar.
2. As an indicator of classification errors. Sequence dissimilarities would indicate that the taxonomist must seriously consider re-assigning classes and species.
3. To provide informative biological markers not found by morphologic examination, permitting more accurate assignment of species.
4. To act as a phylogenetic chronometer to determine when subclasses may have first appeared.
5. To help explain the biological significance of class markers [taxa] that have no apparent scientific explanation (see discussion of Class Unikonta and Class Bikonta in Chapter 15).

The schema for the classification of bacteria that are infectious in humans, appears at the beginning of this chapter. It corresponds closely to the general classification of bacteria proposed by Woese and Fox [21]. Each bacterial subclass will be covered in separate chapters, but there are two divisions of the bacteria that must be discussed here, because they cross taxonomic barriers. These are Gram stainability and G+C content.

A cell wall is a chemical structure that lies outside the membrane that encloses the cytoplasm (i.e., the insides) of bacteria. Some cell walls contain peptidoglycan, and it is this molecule that reacts with the Gram stain to produce a blue color. Some bacteria lack a cell wall, and some bacteria have cell walls that do not contain peptidoglycan. These cells do not react with the Gram stain. Other bacteria contain peptidoglycan in their cell walls, but they have a second, outer membrane that covers the cell wall. Gram stain cannot easily penetrate the outer cell membranes, thus decreasing the Gram stain reaction in these cells. Because there are many different ways by which an organism's Gram reaction is determined, you might guess that Gram staining is a taxonomically perverse property: you would be correct.

With a few exceptions, the Gram-positive cells have a single membrane, and the Gram-negative cells have an inner plus an outer membrane. Most of the Gram-positive cells that are round or rod-shaped fall into Class Bacilli (Chapter 12) or Class Clostridia (Chapter 12). The mollicutes (Chapter 11) are bacteria that have no cell wall to absorb or exclude the Gram stain, but they are traditionally counted among the Gram-positive bacteria because they lack an outer membrane and share a close ancestor with Class Bacilli and Class Clostridia. Among the filamentous bacteria, Class Actinobacteria (Chapter 14) is also Gram-positive.

All bacterial species with an inner and outer membrane are Gram-negative, including all members of Class Proteobacteria (Chapters 5–8). Class Proteobacteria accounts for the majority of Gram-negative pathogens. Other Gram-negative classes are Class Chlamydiae (Chapter 13), Class Spirochaetes (Chapter 9), Class Fusobacteria (Chapter 10), and Class Bacterioidetes (Chapter 10).

Endotoxins are structural molecules found on bacterial cells, that produce generalized inflammatory reactions when injected into humans (e.g., fever, drop in blood pressure, activation of the inflammation and blood cascades). Most endotoxins are lipopolysaccharides found on the outer membrane of bacterial cells; hence, most endotoxins come from Gram-negative bacteria. A classic example of an endotoxin is found in meningococcemia, due to infection with *Neisseria meningitidis* (Beta Proteobacteria, Chapter 6).

The G+C ratio is a measurement of the proportion of Guanine and Cytosine nucleotide bases in an organism's genome. Because all four nucleotide bases (Guanine, Cytosine, Adenine, and Thymidine) are essential constituents of the genetic code, the G+C ratio of most organisms lies in a somewhat restricted range, close to 50%. For certain classes of Gram-positive bacteria, the G+C ratio is used as a useful taxonomic feature. Members of Class Actinobacteria (filamentous cells) have a high G+C ratio, with the ratio of some species exceeding 70%. Members of other Gram-positive classes (Class Bacilli and Class Clostridia) tend to have low G+C ratios (under 50%), though exceptional species of Class Bacilli have G+C ratios exceeding 50% (e.g., *Bacillus thermocatenulatus*). Isolated DNA with a high G+C ratio happens to be more stable than low G+C DNA, and is more resistant to high temperatures. The significance of high G+C content to living organisms, however, is not understood at this time.

Biologists know a great deal about bacterial organisms: their metabolism, biosynthetic capabilities, virulence factors, the sequence of their genomes, how they replicate, and so on (see Glossary item, Virulence factor). Their origin is still somewhat of a mystery. We will re-visit this mystery in Chapter 38 (Overview of Viruses), when we attempt to understand biogenic molecules and their role in the origins of viruses, bacteria, and eukaryotes.

Chapter 5

The Alpha Proteobacteria

"[Bacteria are the] dark matter of the biological world."

Edward O. Wilson

Bacteria
- Proteobacteria
  - Alpha Proteobacteria (Chapter 5)
    - Rhizobiales
      - Bartonellaceae
        - *Bartonella
      - Brucellaceae
        - *Brucella
    - Rickettsiales
      - Anaplasmataceae
        - *Neorickettsia
        - *Ehrlichia
        - *Anaplasma
        - *Wolbachia
      - Rickettsiaceae
        - *Rickettsia
  - Beta Proteobacteria (Chapter 6)
  - Gamma Proteobacteria (Chapter 7)
  - Epsilon Proteobacteria (Chapter 8)
- Spirochaetes (Chapter 9)
- Bacteroidetes (Chapter 10)
- Fusobacteria (Chapter 10)
- Firmicutes
  - Bacilli (Chapter 12)
  - Clostridia (Chapter 12)
  - Mollicutes (Chapter 11)
- Chlamydiae (Chapter 13)
- Actinobacteria (Chapter 14)

J.J. Berman: Taxonomic Guide to Infectious Diseases. DOI: http://dx.doi.org/10.1016/B978-0-12-415895-5.00005-2

Class Proteobacteria contains more human pathogens than any other bacterial class. The human pathogens fall into four biologically distinctive, phylogenetic subclasses (alpha, beta, gamma, and epsilon), discovered through ribosomal RNA sequence analyses [21]. Woese had his own name for Class Proteobacteria; Purple bacteria, based on their descent from a common ancestor that was capable of a photochemical reaction that yielded a photochrome that conferred a purple tinge to bacterial colonies. In addition, all members of Class Proteobacteria are Gram-negative. An outer cell membrane encloses the cell wall and excludes the Gram stain. Class Proteobacteria accounts for the majority of the Gram-negative bacteria.

The Alpha Proteobacteria are characterized by their small size, their Gram negativity, and their intimate associations with eukaryotic cells. The Alpha Proteobacteria live as sybionts, endosymbionts, or as intracellular parasites. This close relationship between Alpha Proteobacteria and Class Eukaryota may extend back to the very first eukaryotic cell. Based on sequence similarities between the Alpha Proteobacteria and eukaryotic mitochondria, it has been proposed that eukaryotic mitochondria evolved from an endosymbiotic member of Class Alpha Proteobacteria.

There are two major subclasses of Class Alpha Proteobacteria: Rhizobiales and Rickettsiales. Class Rhizobiales contains nitrogen-fixing bacteria that live in a symbiotic relationship with plant roots. Without class Rhizobiales, life on earth, as we know it, would cease to exist. Class Rhizobiales contains two human pathogenic genera: Bartonella and Brucella.

Alpha Proteobacteria
  Rhizobiales
    Bartonellaceae
      *Bartonella
    Brucellaceae
      *Brucella

Genus Bartonella, formerly known as Rochalimaea, is a group of facultative intracellular organisms that produce a wide range of diseases, but which seem to share a common life cycle. The pathogenic species of Genus Bartonella are injected into humans from the bite of a blood-feeding vector: fleas, lice, sandflies, and possibly ticks. The Bartonella organism infects the endothelial cells that line blood vessels. Later, the organism leaves the endothelial cell and infects erythrocytes. A blood-feeding vector extracts infected red blood cells from an infected human or from an animal reservoir. The cycle repeats.

The diseases caused by Bartonella species vary, to some extent, on the most favored host cell (endothelial cell, erythrocyte, lymphocyte, monocyte, or granulocyte), on the numbers of infectious organisms, on the chronicity of the infection, and on the pathogenic properties of individual species of the organism (e.g., does it cause red blood cell lysis?). Though Bartonella species can infect healthy persons, most infections arise in immune-compromised patients and in children.

Several of the diseases produced by members of Genus Bartonella are pathologically unique: Carrion's disease, cat-scratch fever, peliosis, and bacillary angiomatosis.

Carrion's disease, caused by *Bartonella bacilliformis*, is endemic in Peru and only occurs in South and Central America. It produces an acute phase characterized by fever, hepatomegaly, splenomegaly, lymphadenopathy, and hemolysis (see Glossary item, Hemolytic syndromes). A chronic phase, known as Verruga Peruana (Peruvian wart) presents as a skin rash characterized histologically by marked proliferation of endothelial cells.

Cat-scratch disease is caused by *Bartonella henselae*, and possibly *Bartonella clarridgeiae*, and has a cat reservoir. The mode of infection is somewhat controversial. Despite the name given to the disease, implying a scratch inoculation, it is likely that fleas carry the bacteria from cats to humans. The disease produces a localized, somewhat persistent, lymphadenopathy, often accompanied by systemic complaints.

Immune-compromised individuals, especially AIDS patients, may form an exaggerated endothelial growth reaction in the skin (bacillary angiomatosis) and in the liver (peliosis hepatis, characterized by localized areas of vascular dilatation and blood pooling). In the skin, the lesions mimic Kaposi's sarcoma (Human herpesvirus 8, in Group I Viruses, Chapter 39), also seen in immune-compromised individuals. In the case of bacillary angiomatosis, Bartonella organisms can be histologically identified in the proliferating endothelial cells. Bacillary angiomatosis can be caused by *Bartonella henselae*, the same organism that causes cat-scratch fever, transmitted by cat scratch, cat bite, and possibly ticks and fleas. Bacillary angiomatosis is also caused by *Bartonella quintana*, transmitted by lice.

Various species of Bartonella have recently been associated with bacteremias with or without endocarditis, myocarditis, and retinitis. It is likely that additional pathogenic Bartonella species will be identified, as the techniques for analyzing blood samples continue to improve.

Readers should be careful not to confuse Bartonella with the similar-sounding Bordetella (Beta Proteobacteria, Chapter 6).

Genus Brucella contains facultative intracellular organisms transmitted from an animal reservoir, usually via drinking unsterilized milk, or through direct contact with animals or their secretions. Human-to-human transmission may occur, rarely. Infection is usually accompanied by fever, arthralgia, myalgia, and bacteremia. Brucella species causing brucellosis include: *Brucella abortus* (cattle reservoir), *Brucella canis* (dog reservoir, rare in humans), *Brucella melitensis* (sheep and cattle reservoir), and *Brucella suis* (pig reservoir, occurring in human handlers).

Because Brucella is a facultative intracellular pathogen, it can live for some time outside of host cells; hence its mode of transmission, via milk and secretions. The ability to survive outside of host cells is unusual among members of Class Alpha Proteobacteria, and has been exploited, in the past, to weaponize strains of Brucella.

Readers should be aware that brucellosis has been known by a great number of different names, including undulant fever (from the undulating, wave-like, progression of fevers), Malta fever, and Mediterranean fever. Mediterranean fever, an arcane synonym for brucellosis, should not be confused with familial Mediterranean fever (a gene disorder characterized by fever and abdominal pain) or with Mediterranean anemia (a synonym for thalassemia).

Alpha Proteobacteria
  Rickettsiales
    Anaplasmataceae
      *Neorickettsia
      *Ehrlichia
      *Anaplasma
      *Wolbachia
    Rickettsiaceae
      *Rickettsia

Class Anaplasmataceae includes four genera containing human pathogens: Neorickettsia, Ehrlichia, Wolbachia, and Anaplasma. All the pathogenic species are intracellular, are morphologically similar to one another, and produce similar diseases.

The disease ehrlichiosis is actually a collection of different infections, caused by organisms in Genus Ehrlichia, Genus Anaplasma, and Genus Neorickettsia, and are vectored from animal reservoirs, by any of several species of ticks (with one exception, *Neorickettsia sennetsu*, transmitted by trematodes). After inoculation into humans, the organisms that cause ehrlichiosis invade and inhabit white blood cells, wherein they can be visualized by microscopic examination. As you would expect from a disease of white cells, the symptoms of ehrlichiosis are systemic, and include headache, fatigue, and muscle aches.

The species producing ehrlichioses include: *Ehrlichia ewingii* (human ewingii ehrlichiosis), *Ehrlichia chaffeensis* (human monocytic ehrlichiosis), and *Ehrlichia canis* (Rickettsia-like infection), *Neorickettsia sennetsu* (sennetsu Ehrlichiosis), and *Anaplasma phagocytophilum* [25]. The last-listed Ehrlichiosis pathogen, *Anaplasma phagocytophilum*, causes human granulocytic anaplasmosis. *Anaplasma phagocytophilum* is the name given to one organism formerly assigned, erroneously, to two different species: *Ehrlichia phagocytophilium* and *Ehrlichia equi* [26]. *Anaplasma phagocytophilium* is endemic to New England and to north-central and north-west United States. *Ehrlichia ewingii* is primarily an infection of deer and dogs. It occurs in humans most commonly in the south central and southeastern states. *Ehrlichia chaffeensis* is most common in the south central and southeastern states. *Ehrlichia canis* is a disease of dogs, with sporadic human cases occurring in the Southeast United States.

Neorickettsia contains one pathogenic species, *Neorickettsia sennetsu*, formerly *Ehrlichia sennetsu*, the cause of Sennetsu ehrlichiosis. The disease is said to mimic a mild case of infectious mononucleosis; both diseases produce a mononucleosis and generalized systemic symptoms. Unique among the Ehrlichioses, *Neorickettsia sennetsu* is transmitted by trematodes, harbored within fish, and eaten undercooked or uncooked, by humans. Readers should be aware that Neorickettsia, despite its name, is not a type of Rickettsia (i.e., not a member of Class Rickettsiaceae). Neorickettsia is a member of Class Anaplasmataceae; hence, the disease it causes is an ehrlichiosis.

Wolbachia contains one, somewhat indirect, human pathogen: *Wolbachia pipientis*.

Wolbachia pipientis happens to be an endosymbiont that infects most members of the filarial Class Onchocercidae [27]. Onchocerca volvulus is the filarial nematode that migrates to the eyes and causes river blindness, the second most common infectious cause of blindness worldwide [28]. *Wolobachia pipientis* lives within *Onchocerca volvulus*, and it is the Wolbachia organism that is responsible for the inflammatory reaction that leads to blindness.

Members of Class Rickettsiaceae are obligate intracellular organisms. Diseases are caused by species in Genus Rickettsia, transmitted by insects, from a reservoir in animals or other humans. The various species of pathogenic rickettsia are split into two groups: the typhus group and the spotted fever group. In the typhus group, disease is louse-borne or flea-borne. In the spotted fever group, disease is usually tick-borne, though some species are transmitted by mites or fleas.

The typhus group contains two pathogenic organisms: *Rickettsia typhi*, the cause of murine typhus or endemic typhus, which is transmitted by fleas that feed on infected rats; and *Rickettsia prowazekii*, so-called epidemic typhus or Brill–Zinsser disease and caused by the human body louse (*Pediculus humanus*), feeding on infected humans. Typhus fever of either type can occur worldwide.

Like so many of the diseases caused by members of Class Alpha Proteobacteria, symptoms are systemic. Typhus is characterized by a high fever. Endemic typhus is usually a milder disease than epidemic typhus. Between 1918 and 1922, epidemic, louse-borne, typhus (*Rickettsia prowazekii*) infected 30 million people, in Eastern Europe and Russia, accounting for about 3 million deaths [29].

The second group of infections, the so-called spotted fever group, is caused by many different rickettsial species. The diseases can be divided into groups based on the transmission vector (i.e., tick, mite, or flea).

Tick-borne:

Rickettsia rickettsii (Rocky Mountain spotted fever, found in Western continents, and not confined to the Rocky Mountains)

Rickettsia conorii (Boutonneuse fever, found in the Mediterranean, Africa, Asia, India)
Rickettsia japonica (Japanese spotted fever, found in Japan)
Rickettsia sibirica (North Asian tick typhus, found in Siberia and China)
Rickettsia australis (Queensland tick typhus, found in Australia)
Rickettsia honei (Flinders Island spotted fever)
Rickettsia africae (African tick bite fever, found in South Africa)
Rickettsia parkeri (American tick bite fever)
Rickettsia aeschlimannii (*Rickettsia aeschlimannii* infection)

Mite-borne:

Rickettsia akari (Rickettsialpox, found in the USA and Russia)
Orientia tsutsugamushi (so-called scrub typhus)

Flea-borne:

Rickettsia felis (Flea-borne spotted fever, found in the Americas, Southern Europe, and Australia)

As the name suggests, the spotted fevers are characterized by rash and fever. Readers should not be confused by the term "scrub typhus" for infection by *Orientia tsutsugamushi* (alternately named *Rickettsia tsutsutgamushi*). This disease is grouped as a "spotted fever," not a form of typhus. As an interesting note, infection by *Orientia tsutsugamushi* has been reported to retard the progression of HIV infection (in terms of viral load) [30]. It seems that the HIV virus shares homologous genes with *Orientia tsutsugamushi*, and these genes produce antigens that are somewhat protective against the virus [30].

Infectious species:

Bartonella bacilliformis (Carrion's disease or Oroya fever or Verruga Peruana or Peruvian wart, bartonellosis)
Bartonella quintana, formerly *Rochalimaea quintana* (trench fever, bacteremia, bacillary angiomatosis, endocarditis)
Bartonella species, formerly Rochalimaea species (bacteremia)
Bartonella henselae (cat scratch disease, bacillary angiomatosis, bacillary peliosis, peliosis hepatis, bacteremia, endocarditis)
Bartonella elizabethae (endocarditis)
Bartonella grahamii (retinitis)
Bartonella vinsoni (endocarditis, bacteremia)
Bartonella washonsis (myocarditis)
Bartonella clarridgiae (bacteremia)
Bartonella rochalimae (Carrion's disease like syndrome)
Brucella abortus (brucellosis or Malta fever)
Brucella canis (dog brucellosis, rarely infecting humans)

Brucella melitensis (sheep and cattle brucellosis, sometimes infecting humans)
Brucella suis (pig brucellosis, transferrable to human handlers)
Neorickettsia sennetsu, formerly *Ehrlichia sennetsu*, formerly *Rickettsia sennetsu* (sennetsu ehrlichiosis)
Anaplasma phagocytophilum, formerly *Ehrlichia phagocytophilium* and *Ehrlichia equi* (human granulocytic anaplasmosis, formerly known as human granulocytic ehrlichiosis)
Ehrlichia ewingii (ewingii ehrlichiosis)
Ehrlichia chaffeensis (which causes human monocytic ehrlichiosis)
Ehrlichia canis (Rikettsia-like infection) [25]
Orientia tsutsugamushi, alternately *Rickettsia tsutsugamushi* (scrub typhus)
Rickettsia typhi, alternately *Rickettsia mooseri* (endemic typhus, murine typhus)
Rickettsia prowazekii (epidemic typhus, Brill–Zinsser disease, flying squirrel typhus)
Rickettsia rickettsii (Rocky Mountain spotted fever)
Rickettsia conorii (Boutonneuse fever)
Rickettsia japonica (Japanese spotted fever)
Rickettsia sibirica (North Asian tick typhus)
Rickettsia australis (Queensland tick typhus)
Rickettsia honei (Flinders Island spotted fever)
Rickettsia africae (African tick bite fever)
Rickettsia parkeri (American tick bite fever)
Rickettsia aeschlimannii (*Rickettsia aeschlimannii* infection)
Rickettsia akari (Rickettsialpox)
Rickettsia felis (Flea-borne spotted fever)
Wolbachia pipientis, the endosymbiont of *Onchocerca volvulus* (river blindness)

Chapter 6

Beta Proteobacteria

"Bacteria will no longer be conceptualized mainly in terms of their morphologies and biochemistries; their relationships to other bacteria will be central to the concept as well."

Carl R. Woese [21]

- Bacteria
  - Preotebacteria
    - Alpha Proteobacteria (Chapter 5)
    - Beta Proteobacteria (Chapter 6)
      - Burkholderiales
        - Alcaligenaceae
          - *Alcaligenes
          - *Bordetella
        - Burkholderiaceae
          - *Burkholderia
      - Neisseriales
        - Neisseriaceae
          - *Eikenella
          - *Neisseria
          - *Kingella
    - Gamma Proteobacteria (Chapter 7)
    - Epsilon Proteobacteria (Chapter 8)
  - Spirochaetes (Chapter 9)
  - Bacteroidetes (Chapter 10)
  - Fusobacteria (Chapter 10)
  - Firmicutes
    - Bacilli (Chapter 12)
    - Clostridia (Chapter 12)
    - Mollicutes (Chapter 11)
  - Chlamydiae (Chapter 13)
  - Actinobacteria (Chapter 14)

J.J. Berman: Taxonomic Guide to Infectious Diseases. DOI: http://dx.doi.org/10.1016/B978-0-12-415895-5.00006-4

Members of Class Beta Proetobacteria are Gram-negative organisms that are either aerobic or facultative anaerobic (i.e., employing respiratory metabolism when oxygen is present and fermentative metabolism when oxygen is absent). Class Beta Proteobacteria has two subclasses with organisms that infect humans: Class Burkholderiales and Class Neisseriales.

Beta Proteobacteria
  Burkholderiales
    Alcaligenaceae
      *Alcaligenes
      *Bordetella
    Burkholderiaceae
      *Burkholderia

Class Burkholderiales contains Class Alcaligenaceae and Class Burkholderiaceae.

Members of Class Alcaligenaceae are found in water and soil and can infect a range of animals. Class Alcaligenaceaa contains two genera with organisms infectious in humans: Alcaligenes and Bordetella.

In humans, Genus Alcaligenes contains opportunistic pathogens that cause disease in immune-compromised patients. *Alcaligenes xylosoxidans* has been found in the respiratory tracts of patients with cystic fibrosis, but its role in respiratory disease is unsettled at this time. *Alcaligenes xylosoxidans* can cause corneal keratitis in patients who use contact lenses [31]. *Alcaligenes faecalis* has been found in some urinary tract infections.

Bordetella species can infect the respiratory tract of healthy individuals. Pertussis, also known as whooping cough, is caused by *Bordetella parapertussis* and *Bordetella pertussis*. *Bordetella bronchiseptica* infects small animals, and rarely infects humans, producing a bronchitis.

Class Burkholderiaceae contains the genus Burkholderia, which includes *Burkholderia mallei*, the cause of glanders, and *Burkholderia pseudomallei*, the cause of melioidosis. Glanders is a serious disease, with a high fatality rate, that is endemic to Africa, Asia, the Middle East, and South America. It is a zoonosis, with a reservoir in horses and other mammals, spread through contaminated water (see Glossary item, Zoonosis). An active surveillance system has eliminated the disease from North America. Infected patients develop lung nodules, upper airway ulcerations, and, eventually, systemic symptoms. Human-to-human transmission occurs, and infection can be transmitted from nasal discharge. Survivors of the infection may become carriers.

Melioidosis, also known as pseudoglanders, is caused by *Burkholderia pseudomallei*. Like glanders, it can have a high mortality. The disease is endemic in Asia and occurs sporadically throughout much of the world. The organism contaminates soil and water; humans are exposed to the bacteria through a break in the skin. As you might expect, rice paddy farmers, who continuously submerse their hands in groundwater, are particularly at risk for exposure. Most humans exposed to the bacteria do not develop disease, but diabetic patients and individuals with chronic diseases are highly susceptible. Incubation may be short (one day) or long (decades). Symptoms and the severity of disease vary widely: fever, pneumonia, and joint pain may occur. Abscesses arising in various organs are a common finding in melioidosis.

The *Burkholderia cepacia* complex is a collection of more than half a dozen related species. *Burkholderia cepacia* is found in soil and water, and produces pneumonia in individuals who have an underlying lung disease (such as cystic fibrosis), or who have weakened immune systems. Infection can be passed from person to person. *Burkholderia cepacia* organisms are difficult to treat (i.e., demonstrate antibiotic resistance) and difficult to remove from the environment (i.e., withstand common disinfectant procedures).

Genus Burkholderia was previously known as Genus Pseudomonas. You may occasionally see species misdesignations in older textbooks.

Beta Proteobacteria
    Neisseriales
        Neisseriaceae
            *Eikenella
            *Neisseria
            *Kingella

Class Neisseriaceae are strictly aerobic and grow chiefly in pairs (i.e., diplococci) or small clusters.

Eikenella corrodens is a normal inhabitant of the mouth, but can produce a bacteremia if mechanically forced into the bloodstream (e.g. chewing bite). It can also act opportunistically in diabetics and immune-compromised individuals. *Eikenella corrodens* can cause cellulitis, as well as endocarditis and is included with the HACEK group of endocarditis-producing organisms (see Glossary item, HACEK).

In Chapter 10, we will describe another oral inhabitant, *Prevotella dentalis* (Class Bacteroidetes), that can cause a bite bacteremia.

Genus Neisseria contains two important pathogens: *Neisseria gonorrhoeae*, the cause of gonorrhea; and *Neisseria meningitidis*, a cause of meningitis. Gonorrhea is a sexually transmitted disease. Approximately 1.5 million new cases of gonorrhea occur annually in North America, where gonorrhea is the third most common sexually transmitted disease [32]. The second most common sexually transmitted disease is chlamydia (Class Chlamydiae, Chapter 13), with about 4 million new cases each year in North America [32]. The most common sexually transmitted disease is trichomoniasis (Class Metamonada, Chapter 16), with about 8 million new cases each year in North America [32].

Neisseria meningitidis is sometimes referred to as meningococcal meningitis, and the organism is often called by an abbreviated form, meningococcus. The bacteremic form of the disease is known as meningococcemia. *Neisseria meningitidis* accounts for about 3000 cases of meningitis in the USA annually. The organism is only found in humans; hence, all infections are thought to be due to human-to-human transmission. Most cases occur in children. The organism is found in a significant percentage of healthy adults (about 10%). Other bacteria cause meningitis in children, including *Hemophilus influenzae* (Gamma Proteobacteria, Chapter 7) and *Streptococcus pneumoniae* (Class Bacilli, Chapter 12); but *Neisseria meningitidis* is the only bacterial organism known to produce epidemics of the disease.

Kingella kingae causes a variety of infectious diseases in children. The organism is commonly found in the throats of children. Diseases include septic arthritis and osteomyelitis. Bacteremia can lead to endocarditis, and *Kingella kingae* is included in the HACEK group of endocarditis-producing organisms (see Glossary item, HACEK).

Infectious species:

Alcaligenes species (corneal keratitis)
Bordetella bronchiseptica (bronchitis)
Bordetella parapertussis (pertussis, also called whooping cough)
Bordetella pertussis (pertussis, also called whooping cough)
Burkholderia cepacia complex (pneumonia in immune-deficient individuals with concurrent lung disease)
Burkholderia mallei, formerly *Pseudomonas mallei* (glanders)
Burkholderia pseudomallei, formerly *Pseudomonas pseudomallei* (melioidosis)
Eikenella corrodens, previously *Bacteroides corrodens* (bite bacteremia, HACEK endocarditis)
Neisseria gonorrhoeae (gonorrhea)
Neisseria meningitidis (meningitis)
Kingella kingae (childhood septic arthritis, childhood osteomyelitis, childhood spondylodiscitis, childhood bacteremia, childhood endocarditis, childhood pneumonia, childhood meningitis, HACEK endocarditis)

Chapter 7

Gamma Proteobacteria

"I learned very early the difference between knowing the name of something and knowing something."

Richard Feynman

Bacteria
  Proteobacteria
    Alpha Proteobacteria (Chapter 5)
    Beta Proteobacteria (Chapter 6)
    Gamma Proteobacteria (Chapter 7)
      Enterobacteriales
        Enterobacteriaceae
          *Edwardsiella
          *Enterobacter
          *Escherichia
          *Klebsiella
          *Morganella
          *Proteus
          *Providencia
          *Salmonella
          *Shigella
          *Yersinia
        Plesiomonaceae
          *Plesiomonas
      Cardiobacteriales
        Cardiobacteriaceae
          *Cardiobacterium
      Legionellales
        Coxiellaceae
          *Coxiella
        Legionellaceae
          *Fluoribacter (formerly Legionella)
          *Legionella

J.J. Berman: Taxonomic Guide to Infectious Diseases. DOI: http://dx.doi.org/10.1016/B978-0-12-415895-5.00007-6

Pasteurellales
Pasteurellaceae
*Aggregatibacter
*Haemophilus
*Pasteurella
Pseudomonadales
Moraxellaceae
*Acinetobacter
*Moraxella
Pseudomonadaceae
*Pseudomonas
Thiotrichales
Francisellaceae
*Francisella
Vibrionales
Vibrionaceae
*Vibrio
Epsilon Proteobacteria (Chapter 8)
Spirochaetes (Chapter 9)
Bacteroidetes (Chapter 10)
Fusobacteria (Chapter 10)
Firmicutes
Bacilli (Chapter 12)
Clostridia (Chapter 12)
Mollicutes (Chapter 11)
Chlamydiae (Chapter 13)
Actinobacteria (Chapter 14)

Members of Class Gamma Proteobacteria share a similarity in their ribosomal RNA, and, like all subclasses of Class Proteobacteria, they are Gram-negative [33]. Otherwise, the Gamma Proteobacteria have widely divergent properties: some are aerobic, others not; some are rods, some are cocci; some have a short curve. others are spiral-shaped; some cells are small, and others are large.

Class Gamma Proteobacteria holds in excess of 20 genera containing about four dozen species that infect humans. It would be unproductive to describe each species in this large and heterogeneous group of organisms. The infectious Gamma Proteobacteria are best understood by understanding the characteristic features of the major subclasses and their genera. The diseases and clinical conditions associated each pathogenic species of the Gamma Proteobacteria are listed at the end of the chapter.

Gamma Proteobacteria
Enterobacteriales
Enterobacteriaceae
*Edwardsiella

*Enterobacter
*Escherichia
*Klebsiella
*Morganella
*Proteus
*Providencia
*Salmonella
*Shigella
*Yersinia
Plesiomonaceae
*Plesiomonas

Class Enterobacteriales contains organisms that live in the intestinal tracts of humans or other organisms. Species in Class Enterobacteriales are commonly known as enterobacteria, and they fall into one of two subclasses: Class Enterobacteriaceae and Class Plesiomonaceae.

Enterobacteria include many organisms; some are aerobic, others are anaerobic; most are facultative anaerobes. Many enterobacteria are non-pathogenic commensals, that live in the human intestines, becoming pathogenic under abnormal circumstances. Other enterobacteria are always pathogenic. The enterobacteria are rod-shaped. Most are about 1–5 microns in length.

Edwardsiella contains several species that live in a variety of animals, including snakes, seals, and fish. A rare disease, Edwardsiellosis typically occurs after ingestion of insufficiently cooked freshwater fish (which harbors the organism in its GI tract). Most cases occur in tropical climates (e.g., Southeast Asia, India, Cuba). A case possibly contracted from ornamental (aquarium) fish has been reported [34]. The disease usually presents in humans as a gastroenteritis which may be severe, and occasionally fatal.

Enterobacter contains rod-shaped species that live in the human intestines. Several enterobacters are opportunistic pathogens, including *Enterobacter aerogenes* and *Enterobacter cloacae*. Disease caused by Enterobacter species usually involves the urinary tract and the lungs.

Escherichia is another rod-shaped factultatively anaerobic genus with species that inhabit the human intestine. Most species of Escherichia are non-pathogenic commensals. Other species cause enteritis or urinary tract infections. *Escherichia coli* has pathogenic and non-pathogenic strains, and is the Escherichia species most commonly responsible for human disease. One strain of *Escherichia coli* (strain O157:H7) produces a toxin (shiga-like toxin) that confers heightened pathogenicity, producing a hemorrhagic enteritis. This dangerous strain of *Escherichia coli* is transmitted to humans, often by ingesting contaminated beef, water, or vegetables that have not been properly sanitized.

Genus Klebsiella contains several species of rod-shaped bacteria with thick polysaccharide capsules. Species are found throughout the environment.

Klebsiella pneumoniae, the most common pathogen in humans, inhabits the gastrointestinal tracts of diverse animals, either chronically or transiently, and is found in sewage and soil [35]. *Klebsiella pneumoniae*, as its name suggests, causes pneumonia in humans. *Klebsiella pneumoniae* and its closely related relative, *Klebsiella oxytoca*, are capable of producing a wide range of conditions in humans, particularly sepsis and urinary tract infections.

Klebsiella granulomatis, formerly *Calymmatobacterium granulomatis*, formerly *Donovania granulomatis*, is the cause of granuloma inguinale, also known as donovanosis, also known as granuloma venereum. Infection produces genital ulcers, and can be mistaken clinically with two other diseases that are characterized by genital ulcers: syphilis (Class Spirochaetae, Chapter 9), and chancroid (see below). In addition, granuloma venereum, caused by *Klebsiella granulomatis*, must not be confused with lymphogranuloma venereum, caused by *Chlamydia trachomatis* (Class Chlamydiae, Chapter 13).

Rhinoscleroma is a granulomatous disease, endemic to Africa, Southeast Asia, South America, and some parts of Eastern Europe. Cases may occur in the USA. The nose is involved by rhinoscleroma in over 95% of cases.

Readers should not confuse rhinscleroma, caused by *Klebsiella rhinoscleromatis* (Class Gamma Proteobacteria, Chapter 7), with rhinosporidiosis, caused by *Rhinosporidium seeberi* (Choanozoa, Chapter 23).

Morganella is a genus of anaerobic organisms, only one of which is infectious in humans: *Morganella morganii*. This organism lives in the intestines of animals (including humans). It can cause a wide variety of diseases, including enteritis (causing so-called summer diarrhea), sepsis, and organ infections. The bacterial genus Morganella should not be confused with the fungal genus of the same name.

Genus Proteus contains rod-shaped bacteria that inhabit soil and water. Species in the genus produce urease. The genus contains three opportunistic pathogens that are sometimes found in the human intestinal tract: *Proteus mirabilus*, *Proteus vulgaris*, and *Proteus penneri*. All of the Proteus species are capable of causing urinary tract infections. The species most commonly found in Proteus-caused infections is *Proteus mirabilis*.

Proteus species have a particular association with struvite stones of the kidney. Because Proteus species produce urease, the resulting high levels of ammonia in urine leads to alkalinity, which in turn favors the crystallization of struvite (an ammonia-containing mineral) in urine. Kidney stones passed in urine, or surgically extracted, are routinely analyzed for chemical composition. Stones composed of struvite are uncommon. Most stones contain calcium oxalate, calcium phosphate, or uric acid. A struvite stone should alert clinicians that the patient may have a Proteus urinary tract infection.

Genus Providencia contains species that are found in soil, water, and sewage. In humans, Providencia species are opportunistic pathogens, superinfecting burns or causing urinary tract infections in patients with indwelling catheters. *Providencia stuartii* is a common cause of so-called purple urine bag syndrome, a condition associated with various infectious members of Class

Proteobacteria that cause urinary tract infections in catheterized patients (i.e., patients whose urine is caught in a plastic bag). An early name given to members of Class Proteobacteria was "purple bacteria," based on their purple tinge. Urine contaminated with purple bacteria causes purple urine bag syndrome, and the most common proteobacteria causing the syndrome is *Providencia stuartii*.

Members of Genus Salmonella are found in a wide variety of animals, including reptiles, and pathogenic species produce enteritis, often transmitted through contaminated food (i.e., so-called salmonellosis food poisoning). Some taxonomists list the pathogenic species as serovars under the name *Salmonella enterica enterica*.

Salmonella typhi and *Salmonella paratyphi* cause typhoid and paratyphoid, respectively. Typhoid, also known as typhoid fever, is a severe gastroenteritis, usually accompanied by a high fever. Paratyphoid is a similar, but generally less severe disease. Neither of these diseases should be confused with typhus fever, caused by *Rickettsia typhi* and *Rickettsia prowazekii*, and transmitted by fleas and body lice (respectively). Both diseases (typhoid and typhus) take their root from a Greek word meaning stupor, referring to the neurologic manifestations of the diseases.

Species of Genus Shigella cause bacillary dysentery, also known as shigellosis. Despite its disease name, readers should not assume that the cause of bacillary dysentery is a member of Class Bacilli (Chapter 12). The exclusive cause of bacillary dysentery is Shigella species (although a sister species, *Yersinia enterocolitica*, is sometimes included).

Shigella boydii (Gamma Proteobacteria, Chapter 7), one of the causes of shigellosis, should not be confused with *Pseudallescheria boydii*, a fungus in Class Ascomycota (Chapter 46), one of many fungal organisms associated with the skin infection maduromycosis.

Yersinia enterocolitica is the cause of yersiniosis, a type of enteritis [36]. It is a zoonotic infection found in a wide variety of animals, including cattle. Humans are most often infected by ingesting contaminated food or water. Genus Yersinia contains species that are siderophilic (iron-loving). Consequently *Yersinia enterocolitica* seems to have a high infection rate in patients who have a high blood heme content (e.g., hemochromatosis patients). In addition, *Yersinia enterocolitica* seems to thrive in stored blood and is occasionally found as a contaminant of blood products.

Plague, caused by *Yersinia pestis*, is not an extinct diseases. Each year, several thousand cases of plague occur worldwide, resulting in several hundred deaths. Virtually all of the cases occur in Africa. Its reservoir is small animals (e.g., rodents), and its most common route of spread is from animal to flea to animal. Humans can be infected by flea bite, by direct or indirect contact with infected animals, or by aerosolized droplets produced by infected humans. The clinical manifestation of plague is influenced by the mode of infection: flea bites lead to the bubonic form, characterized by large, necrotic lymph nodes (buboes); entry through skin abrasions, leading to blood infection, may cause a septicemic form of plague; and infection by aerosolized droplets may produce the pneumonic form of plague.

Class Plesiomonaceae is the second subclass contained in Class Enterobacteriales. It contains one genus, with one currently recognized human pathogenic species: *Plesiomonas shigelloides*, an organism found in diverse animals, including fish and shellfish. *Plesiomonas shigelloides* was thought to be an exclusively opportunistic pathogen, producing a gastroenterities, capable of developing into sepsis, in immune-compromised patients. It has now been found to be associated with a significant number of cases of mild enteritis in otherwise healthy individuals [37].

Readers should not confuse *Plesiomonas shigelloides*, containing the species name "shigelloides," with the genus name "Shigella" (see above).

Gamma Proteobacteria
  Cardiobacteriales
    Cardiobacteriaceae
      *Cardiobacterium

Genus Cardiobacterium contains one pathogenic species: Cardiobacterium hominis, a pleomorphic but often rod-shaped bacteria that causes some cases of endocarditis. It is included with the HACEK organisms. See Glossary item HACEK.

Gamma Proteobacteria
  Legionellales
    Coxiellaceae
      *Coxiella
    Legionellaceae
      *Legionella
      *Fluoribacter (formerly Legionella)

Whereas most subclasses of the Gamma Proteobacteria live extracellular lives, the members of Class Legionellales that infect humans live inside cells.

Genus Coxiella contains one human pathogen: *Coxiella burnetii*, the cause of Q fever. *Coxiella burnetii* is an obligate intracellular organism that is capable of reverting to a small cell variant of itself, that is resistant to intracellular degradation by lysosomes. This highly infectious zoonosis is transmitted by inhaling aerosolized respiratory material, or from direct contact with secretions from an infected animal (including cattle, sheep, goats, dogs, and cats). A flu-like syndrome results, which may progress to chronic pneumonia, ARDS (acute respiratory distress syndrome) and various systemic sequelae, including endocarditis.

Genera of Class Legionellaceae are facultative intracellular organisms and include Legionella and Fluoribacter [38].

Legionella species live within amoeba in the environment. Infection occurs after inhalation of the bacteria, and epidemics of Legionnaire's disease have been linked to contaminated sources of aerosolized water, and from water-holding systems. The names of the disease and the organism derive from the first diagnosed epidemic, occurring in members of an American Legion who attended a bicentennial convention in Philadelphia, in July, 1976. Direct person-to-person spread has not been established. Disease most often occurs in immune-compromised individuals and the elderly. Infection is usually pulmonary and can be fatal. Pontiac fever is a milder form of Legionnaire's disease. Between 10 000 and 50 000 cases of Legionnaire's disease occur each year in the United States. The bacteria can be visualized in tissue sections with a silver stain. Aside from *Legionella pneumophila*, there are more than 50 species of Legionella, some of which have been shown to produce Legionnaire's disease.

DNA analysis of species of Legionella demonstrated sequence dissimilarities, suggesting that Genus Legionella could be assigned several genera [39]. Genus Fluoribacter was created as a new genus in Class Legionellaceae, and this Genus was assigned *Fluoribacter bozemanae*, formerly *Legionella bozemanae*. The species in Genus Fluoribacter behave clinically like Legionella species.

Gamma Proteobacteria
  Pasteurellales
    Pasteurellaceae
      *Aggregatibacter
      *Haemophilus
      *Pasteurella

Class Pasteurellaceae contains numerous species of factultative anaerobic bacteria, that are, in most cases, rod-shaped. Most species live as commensals in vertebrates, and are found primarily in the upper respiratory tracts, reproductive tracts and, in some cases, the gastrointestinal tract. Three genera contain species pathogenic in humans: Aggregatibacter, Haemophilus, and Pasteurella.

Genus Aggregatibacter contains the oral commensal *Aggregatibacter actinomycetemcomitans*, formerly *Actinobacillus actinomycetemcomitans*, which causes an aggressive form of periodontitis.

Genus Haemophilus contains the pathogenic species: *Haemophilus influenzae*, *Haemophilus parainfluenzae*, and *Haemophilus ducreyi*.

Haemophilus influenzae is found in the upper respiratory tract of normal individuals. It causes pneumonitis, meningitis, conjunctivitis, otitis media, and bacteremia, in infants and young children. An unencapsulated strain causes conjunctivitis and otitis media. Its species name, influenzae, was assigned when the bacteria was mistakenly thought to be the cause of influenza. Influenza, also known as the flu, is caused exclusively by the influenza virus, a Group V Orthomyxovirus (Chapter 43).

Haemophilus parainfluenzae causes some cases of endocarditis. It is included among the HACEK organisms (see Glossary item HACEK).

Readers should remember that despite its name, *Haemophilus parainfluenzae* is not the cause of the disease known as parainfluenza. Parainfluenza is a type of croup (laryngotracheobronchitis), and about 75% of the cases of croup are caused by the parainfluenza virus, a Group V virus (Chapter 43).

Haemophilus ducreyi is the cause of chancroid, a sexually transmitted disease. Chancroid produces painful genital ulcers. It must not be confused with other diseases that cause genital ulcers: syphilis and granuloma inguinale. *Klebsiella granulomatis*, in Class Enterobacteriaceae causes granuloma inguinale, and was described previously in this chapter. Syphilis is caused by a spirochete (Chapter 9). Adding to the confusion, the syphilitic genital ulcer is known as a chancre.

Genus Pasteurella contains *Pasteurella multocida*, the cause of pasteurellosis, a zoonosis. *Pasteurella multocida* lives as a commensal or as a pathogen in a variety of animals. Infection in humans usually results from close contact with infected animals. Infections may arise through bites or scratches from pets, particularly cats. Infection can result in sepsis, pneumonia, or skin infections. Skin infections often occur at the site of a cat scratch.

Gamma Proteobacteria
    Pseudomonadales
        Moraxellaceae
            *Acinetobacter
            *Moraxella
        Pseudomonadaceae
            *Pseudomonas

Class Pseudomonadales contains organisms that cause opportunistic infections. Several grow as surface biofilms, making them hard to disinfect from contaminated medical devices, and thus potential sources of nosocomial (hospital-acquired) infections. Biofilm infections in hospitals will be described again when we discuss the gliding bacteria (Chapter 10).

Members of Genus Acinetobacter are found in soil. The organisms grow as coccobacillary forms, sometimes clumped in pairs. Species of Acinetobacter are found in hospitals, particularly as surface colonies. Though Acinetobacter species are generally non-pathogenic in healthy individuals, serious infections may occur in weakened individuals in intensive care units, on ventilatory support, or recovering from wounds. Outbreaks of Acinetobacter infections have occurred in military hospitals. *Acinetobacter baumanii* is often associated with pneumonia in hospital patients, particularly in patients receiving ventilator support. Infections may lead to bacteremia, meningitis, and skin infections. Bacteria and disseminated infections are often associated with another species, *Acinetobacter lwoffi*.

Readers should not confuse Genus Acinetobacter (Gamma Proteobacteria, Chapter 7), with Class Actinobacteria (Chapter 14).

Moraxella catarrhalis is an opportunistic pathogen that often targets young children, the elderly, and immune-compromised individuals. It can cause a wide range of infections. In adults it tends to cause pneumonia, bronchitis, sinusitis, and can lead to sepsis. In young children, it can cause otitis media. The organism seems to live in humans exclusively, colonizing the respiratory tracts of infected individuals [40].

Much like Genus Acinetobacter, members of Genus Pseudomonas are opportunistic infections that are often nosocomial (i.e., hospital-acquired). Pseudomonas species can grow as biofilms. The biofilms contain an excreted exopolysaccharide that interferes with disinfection of colonized surfaces. Patients in a weakened condition are particularly vulnerable to infections. The most common Pseudomonas species involved in human infections is *Pseudomonas aeruginosa*, which causes some of the same kinds of infections observed with Acinetobacter species (pneumonia, bacteremia, urinary tract infections, burn infections).

Several species, formerly assigned to Genus Pseudomonas, are now assigned to Genus Burkholderia (Beta Proteobacteria, Chapter 6). These include species now known as *Burkholderia mallei* and *Burkholderia pseudomallei*.

Gamma Proteobacteria
  Thiotrichales
    Francisellaceae
      *Francisella

Class Thiotrichales has received scientific attention as it contains the largest known species of bacteria, *Thiomargarita namibiensis*. This ocean-dwelling non-pathogen can reach a size of 0.75 millimeter, visible to the unaided eye (see Glossary item, Largest species).

In contrast, Genus Francisella contains small organisms, under a micron in size. *Francisella tularensis* is the cause of tularemia, also known as rabbit fever. The species occurs in two serotypes (A and B), with Type B producing milder disease. The organism's reservoir is infected rodents (including rabbits and beavers), and is usually transmitted to humans via arthropod bites, particularly tick bites. Human infections can also follow ingestion of contaminated water or soil or by handling infected animals. Tularemia is a serious disease, that can produce a wide variety of symptoms. A common presentation involves lymph nodes (glandular form) and infection of the facial skin. The enlargement of lymph nodes is similar to the buboes (ulcerating enlarged lymph nodes) associated with plague. Other species of Genus Francisella, *Francisella novicida* and *Francisella philomiragia*, previously assigned to Genus Yersinia, can produce sepsis.

Gamma Proteobacteria
  Vibrionales
    Vibrionaceae
      *Vibrio

Genus Vibrio contains motile species characterized by a curved shape and a distinctive polar flagellum. Many Vibrio species are pathogens for fish and crustaceans. *Vibrio cholerae* is the cause of cholera, a particularly aggressive enteritis. Sporadic cholera cases, in developed countries, most often occur after ingesting undercooked seafood. Large outbreaks of cholera usually result from poor sanitary conditions, with infected humans contaminating the water supply. Worldwide, about 3 to 5 million people contract cholera, annually. Other Vibrio species that are pathogenic in humans are *Vibrio parahaemolyticus* and *Vibrio vulnificus*, both of which cause enteritis.

Infectious species:

Edwardsiella tarda (Edwardsiellosis)
Enterobacter aerogenes (opportunistic sepsis)
Enterobacter cloacae (urinary tract infections, respiratory tract infections)
Escherichia coli (food poisoning, enteritis)
Klebsiella oxytoca (colitis, sepsis)
Klebsiella pneumoniae (pneumonia)
Klebsiella rhinoscleromatis (rhinoscleroma)
Klebsiella granulomatis, formerly *Calymmatobacterium granulomatis*, formerly *Donovania granulomatis* (granuloma inguinale, donovanosis, granuloma venereum)
Morganella morganii (a wide variety of organ infections, and sepsis)
Plesiomonas shigelloides (gastroenteritis, developing into sepsis, in immune-compromised humans)
Proteus mirabilis (struvite renal stones, urinary tract infections)
Proteus penneri, formerly a substrain of *Proteus vulgaris* (nosocomial urinary tract infections and sepsis)
Proteus vulgaris (nosocomial urinary tract infections and sepsis)
Providencia species (urinary tract infections, gastroenteritis, and bacteremia)
Providencia stuartii (purple urine bag syndrome)
Salmonella arizonae (salmonellosis)
Salmonella enteritidis (gastroenteritis)
Salmonella paratyphi A, B, C (paratyphoid fever)
Salmonella typhi, Salmonella enterica enterica, serovar Typhi (typhoid fever)
Salmonella typhimurium (food poisoning)
Salmonella choleraesuis (food poisoning)
Shigella boydii (shigellosis, bacillary dysentery)
Shigella dysenteriae (shigellosis, bacillary dysentery)
Shigella flexneri (shigellosis, bacillary dysentery)
Shigella sonnei (shigellosis, bacillary dysentery)
Yersinia enterocolitica (yersiniosis, occasionally included with Shigella species as a cause of bacillary dysentery)
Yersinia pestis (plague)

Yersinia pseudotuberculosis (pseudotuberculosis)
Cardiobacterium hominis (endocarditis)
Coxiella burnetii (Q fever)
Fluoribacter bozemanae, formerly *Legionella bozemanae* (pneumonia)
Legionella pneumophila (Legionellosis, Legionnaire's disease, Legion fever, Pontiac fever)
Legionella species (Legionellosis, Legionnaire's disease, Legion fever, Pontiac fever)
Haemophilus ducreyi (chancroid)
Haemophilus influenzae (meningitis, pneumonia, conjunctivitis, otitis media)
Haemophilus parainfluenzae (endocarditis)
Pasteurella multocida (pasteurellosis)
Acinetobacter baumannii (nosocomial pneumonia, ventilator-associated pneumonia)
Acinetobacter lwoffi (nosocomial pneumonia, nosocomial bacteremia, nosocomial meningitis, nosocomial wound infections)
Moraxella catarrhalis (respiratory tract infections, sepsis, especially in immune-compromised individuals)
Pseudomonas aeruginosa (multi-organ infections, sepsis in immune-compromised individuals, hot tub rash)
Francisella tularensis (tularemia, rabbit fever)
Vibrio cholerae, including El Tor (cholera)
Vibrio parahaemolyticus (watery diarrhea)
Vibrio vulnificus (watery diarrhea)
Aggregatibacter actinomycetemcomitans, formerly *Actinobacillus actinomycetemcomitans* (periodontal disease)

Chapter 8

Epsilon Proteobacteria

"Names are an important key to what a society values. Anthropologists recognize naming as one of the chief methods for imposing order on perception."

David S. Slawson

Bacteria
  Proteobacteria
    Alpha Proteobacteria (Chapter 5)
    Beta Proteobacteria (Chapter 6)
    Gamma Proteobacteria (Chapter 7)
    Epsilon Proteobacteria (Chapter 8)
      Campylobacterales
        Helicobacteraceae
          *Helicobacter
        Campylobacteraceae
          *Campylobacter
  Spirochaetes (Chapter 9)
  Bacteroidetes (Chapter 10)
  Fusobacteria (Chapter 10)
  Firmicutes
    Bacilli (Chapter 12)
    Clostridia (Chapter 12)
    Mollicutes (Chapter 11)
  Chlamydiae (Chapter 13)
  Actinobacteria (Chapter 14)

Astute readers may notice that there is no chapter for the Delta Proteobacteria. This is because the Delta Proteobacteria contain no organisms that are pathogenic in humans. Circumstances may change; Class Delta Proteobacteria, Genus Desulfovibrio has been associated with an opportunistic human infection [41]. Until more definitive studies come to light, we can skip directly to Class Epsilon Proteobacteria.

J.J. Berman: Taxonomic Guide to Infectious Diseases. DOI: http://dx.doi.org/10.1016/B978-0-12-415895-5.00008-8

Epsilon Proteobacteria, like all other proteobacteria, are Gram negative bacteria, with an inner and outer membrane enclosing the cell wall. These bacteria are curved or spiral-shaped and most genera live in the digestive tracts of animals. Two genera are pathogenic in humans: Campylobacter and Helicobacter. Helicobacter species live in the stomach. Campylobacter species live in the duodenum.

Epsilon Proteobacteria
  Campylobacterales
    Helicobacteraceae
      *Helicobacter
    Campylobacteraceae
      *Campylobacter

More than half of the world's human population is infected by *Helicobacter pylori*. Infection rates are highest in developing countries. *Helicobacter pylori* lives in the stomach, an organ that, for many years, was thought to be sterile: no bacteria could possibly survive the acidic environment of the gastric lining! In 1983, Robin Warren and Barry Marshall described what they called "unidentified curved bacilli" in patients with gastritis, by direct visualization under a microscope. These structures were later proved to be *Helicobacter pylori*. To demonstrate the viability and the pathogenicity of these organisms, Marshall experimented on himself, by ingesting gastric juice from a "dyspeptic" man. About ten days later, he developed gastritis, suggesting that an agent in stomach contents from patients with gastritis, could transmit the disease to other people. Marshall and Warren submitted their findings in a scientific abstract, which was rejected by the Australian Gastroenterological Association [42]. Luckily for us, these two stalwart scientists persevered. In 2005, the Nobel prize for medicine was awarded to Robin Warren and Barry Marshall for discovering the role of *Helicobacter pylori* in gastric disease. In retrospect, we now know that these bacteria were observed in histologic examination of gastritis specimens as early as 1875 [43]. Until Warren and Marshall, nobody made the cognitive leap, connecting the visualized organisms to a human disease; pathologists were intellectually invested in the false belief that the stomach was a sterile organ.

The mechanism of transmission of Helicobacter has not been conclusively determined, but the organism has been isolated from feces, saliva, and tooth plaque from infected individuals, suggesting a direct human-to-human spread [44]. Infections tend to be persistent. *Helicobacter pylori* is believed to be a common cause of peptic ulcers, chronic gastritis, duodenitis, and stomach cancers; both adenocarcinoma of stomach and MALT (mucosa-associated lymphoid tissue) lymphoma. Since the discovery of *Helicobacter pylori*, a variety of additional Helicobacter species have been isolated from animals and from humans; many are enteropathogenic. Examples are *Helicobacter suis*, *Helicobacter felis*, *Helicobacter bizzozeronii*, and *Helicobacter salomonis*.

Some of these newly discovered strains seem to be have greater disease-causing potential than *Helicobacter pylori*, with more organs involved. *Helicobacter cinaedi* causes some cases of cellulitis in immune-compromised individuals [45]. A role for Helicobacter species in Crohn's disease and other enteric and hepatic inflammatory conditions is currently being studied. Numerous strains have been detected by PCR (polymerase chain reaction), but not isolated from tissues.

The second Genus of Class Epsilon Proteobacteria is Campylobacter, the cause of campylobacteriosis. About a dozen species of Campylobacter are currently under study as possible causes of human disease. *Campylobacter jejuni* lives in the small intestine and colon, and infections typically cause an acute, self-limited enteritis. *Campylobacter fetus* is an opportunistic pathogen that causes sepsis in newborns.

Infectious species:

Campylobacter fetus (sepsis in newborns)
Campylobacter jejuni (enteritis)
Helicobacter pylori, formerly *Campylobacter pylori*, formerly *C. pyloridis* (gastritis, increased risk of gastric cancer and gastric MALT lymphoma)
Helicobacter cinaedi (cellulitis) [45]

Chapter 9

Spirochaetes

"And what physicians say about disease is applicable here: that at the beginning a disease is easy to cure but difficult to diagnose; but as time passes, not having been treated or recognized at the outset, it becomes easy to diagnose but difficult to cure."

Niccolo Machiavelli

Bacteria
- Proteobacteria (Chapters 5–8)
  - Alpha Proteobacteria (Chapter 5)
  - Beta Proteobacteria (Chapter 6)
  - Gamma Proteobacteria (Chapter 7)
  - Epsilon Proteobacteria (Chapter 8)
- Spirochaetes (Chapter 9)
  - Spirochaetales
    - Leptospiraceae
      - *Leptospira
    - Spirochaetaceae
      - *Borrelia
      - *Treponema
    - Brachyspiraceae (formerly Serpulinaceae)
      - *Serpulina
      - *Brachyspira
    - Spirillaceae
      - *Spirillum
- Bacteroidetes (Chapter 10)
- Fusobacteria (Chapter 10)
- Firmicutes
  - Bacilli (Chapter 12)
  - Clostridia (Chapter 12)
  - Mollicutes (Chapter 11)
- Chlamydiae (Chapter 13)
- Actinobacteria (Chapter 14)

J.J. Berman: Taxonomic Guide to Infectious Diseases. DOI: http://dx.doi.org/10.1016/B978-0-12-415895-5.00009-X

It was not too long ago that microbiologists classified spirochetes among the protozoa [46]. We now know that the spirochetes are bacteria with a peculiar morphologic feature that is characteristic of every member of the class. As mentioned in the overview chapter (Chapter 4), Class Spirochaetes are Gram-negative organisms, and have an inner and outer membrane. Spirochetes are long, helical organisms with an axial flagellum that runs between the inner and the outer membranes. In this location, motion of the flagellum twists the entire organism, and this twisting motion accounts for the motility of the organism.

Most of the members of Class Spirochaetes require direct transmission from an animal reservoir, either through a bite of a louse (Class Hexapoda), tick (Class Chelicerata), rat (Class Craniata), through intimate sexual contact with an infected human, or through close skin-to-skin contact with another human (e.g. yaws). The exception is Genus Leptospira.

Spirochaetes
  Spirochaetales
    Leptospiraceae
      *Leptospira
    Spirochaetaceae
      *Borrelia
      *Treponema
    Brachyspiraceae (formerly Serpulinaceae)
      *Serpulina
      *Brachyspira
    Spirillaceae
      *Spirillum

Numerous species of Genus Leptospira can cause leptospirosis; also known as Weil disease, Weil syndrome, canicola fever, canefield fever, nanukayami fever, 7-day fever, rat catcher's yellows, Fort Bragg fever, and pretibial fever. The disease produces a bacteremia, with consequent splenitis and infection of multiple organs. Jaundice is a common feature of the disease. Leptospirosis is transmitted to humans through contact with the body fluids of infected animals. Drinking water or eating food that has been contaminated with infected urine is a common route of transmission. A wide range of animals may carry the infection, but rats and mice are usually considered the most important hosts. Though leptospirosis is a rare disease, it can occur just about anywhere in the world, often as localized outbreaks. Nepal has been a frequent site of leptospirosis outbreaks.

Borrelia burgdorferi causes Lyme disease. It is transmitted by the bite of infected deer ticks. The rates of Lyme disease infections are increasing in the USA. Though the disease my occur anywhere, it is found most often in the Northern hemisphere. In Europe, most cases of Lyme disease are caused by *Borrelia afzelii* and by *Borrelia garinii*. It is likely that newly identified species of Borrelia, capable of producing Lyme disease, will be encountered in the near future. Lyme disease has been known to produce a bewildering array of symptoms, particularly when it enters its chronic stage. Thus, it has been confused with various chronic ailments and neurodegenerative disorders of unknown etiology.

Relapsing fever is an infection characterized by recurring fevers often accompanied by other systemic symptoms including fever, nausea, and rash. It is caused by any of several genera of class Borrelia and is transmitted by a tick or a louse. Tick-borne relapsing fever (TBRF) is transmitted by the Ornithodoros tick, a genus in Class Argasidae, whose members are all soft-bodied. It occurs in many areas of the world and is found in the United States and Canada. Species associated with tick-borne relapsing fever are: *Borrelia duttoni*, *Borrelia parkerii*, and *Borrelia hermsii*. Louse-borne relapsing fever is caused by *Borrelia recurrentis* and occurs most often in Africa, Asia, and Latin America.

Borrelia lonestari (Southern tick-associated rash illness or STARI) is probably the most redundant name in taxonomy, as it encompasses the abbreviation STARI (loneSTARI) and its common geographic location (Texas, the Lone Star state). Though the disease occurs in Texas, it can also be found in every state between Texas and Maine.

Brachyspira pilosicoli, formerly *Serpulina pilosicoli*, is a colonizing spirochete in the large intestine of pigs, and is occasionally found in humans. It can produce diarrhea and rectal bleeding [47]. *Brachyspira aalborgi* accounts for a small fraction (less than 1%) of cases of acute appendicitis.

Subspecies of *Treponema pallidum* are responsible for four diseases: syphilis, pinta, bejel, and yaws. Syphilis, caused by *Treponema pallidum* (pallidum variant), is transmitted sexually. It can also be transmitted from mother to infant via transplacental infection or through physical exposure to the neonate during its passage through the birth canal. Humans are the only animal reservoir for syphilis. In adults, syphilis is a multistage disease, first appearing as a primary skin lesion (chancre), followed weeks later by a rash and various systemic symptoms (secondary syphilis), followed by a long latency period (years), followed by a systemic illness affecting the brain, cardiovascular system, and other organs (tertiary syphilis). The tertiary phase is non-infectious. Today, syphilis is most common in developing countries [48].

Pinta, bejel, and yaws are non-venereal infections caused by subspecies of *Treponema pallidum* (variants carateum, endemicum, and pertenue, respectively). Each is primarily a skin disease that is spread by direct skin-to-skin contact. Lesions beginning in the skin can spread to joints and bones. Pinta occurs primarily in Central and South America. Yaws occurs primarily in Asia, Africa, and South America. Most cases of bejel occur in the Mediterranean region and Northern Africa and often involve the mouth and oral mucosa, with mouth-to-mouth transmission.

Treponema denticola is a major cause of periodontitis.

The taxonomic assignment of genus Spirillum is currently unstable, being placed in either Class Spirochaetes or in Class Beta Proteobacteria. Adding to the confusion, the only human pathogen in the genus, produces a disease that is attributed to two separate classes of organisms. Rat-bite fever is caused by either *Spirillum minus* or by *Streptobacillus moniliformus* (Class Fusobacteria, Chapter 10) [49]. Evidently, diseases do not read taxonomic textbooks. Regardless of the causative organism, or the phylogenetic classes to which the organisms are assigned, the clinical symptoms are similar, as is the treatment.

Infectious species:

Leptospira species, including *Leptospira interrogans* (leptospirosis, Weil disease, Weil syndrome, canicola fever, canefield fever, nanukayami fever, 7-day fever, rat catcher's yellows, Fort Bragg fever, and pretibial fever)
Borrelia burgdorferi (Lyme disease in the United States)
Borrelia afzelii (Lyme disease in Europe)
Borrelia garinii (Lyme disease in Europe)
Borrelia duttonii (tick-borne relapsing fever, typhinia)
Borrelia recurrentis (louse-borne relapsing fever, typhinia)
Borrelia hermsii (tick-borne relapsing fever, typhinia)
Borrelia parkerii (tick-borne relapsing fever, typhinia)
Borrelia lonestari (Southern tick-associated rash illness, STARI)
Brachyspira pilosicoli, formerly *Serpulina pilosicoli* (diarrhea, spirochetosis)
Brachyspira aalborgi (appendicitis)
Treponema carateum, alternately known as *Treponema pallidum* var carateum (pinta)
Treponema pallidum, alternately known as *Treponoma pallidum* var pallidum (syphilis, congenital syphilis)
Treponema endemicum, alternately known as *Treponema pallidum* var endemicum (bejel)
Treponema pertenue, alternately known as *Treponema pallidum* var pertenue (Yaws, frambesia tropica, Pian)
Treponema denticola (periodontal disease and opportunistic infections)
Spirillum minus (rat-bite fever)

Chapter 10

Bacteroidetes and Fusobacteria

"The greatest enemy of knowledge is not ignorance, it is the illusion of knowledge."
Stephen Hawking

Bacteria
  Proteobacteria
    Alpha Proteobacteria (Chapter 5)
    Beta Proteobacteria (Chapter 6)
    Gamma Proteobacteria (Chapter 7)
    Epsilon Proteobacteria (Chapter 8)
  Spirochaetes (Chapter 9)
  Bacteroidetes (Chapter 10)
    Bacteroidales
      Bacteroidaceae
        *Bacteroides
      Porphyromonadaceae
        *Porphyromonas
      Prevotellaceae
        *Prevotella
    Flavobacteria
      Flavobacteriales
        Flavobacteriaceae
          *Elizabethkingia
  Fusobacteria (Chapter 10)
    Fusobacteriales
      Fusobacteriaceae
        *Fusobacterium
        *Streptobacillus
  Firmicutes (low G+C Gram+)
    Bacilli (Chapter 12)
    Clostridia (Chapter 12)
    Mollicutes (Chapter 11)

J.J. Berman: Taxonomic Guide to Infectious Diseases. DOI: http://dx.doi.org/10.1016/B978-0-12-415895-5.00010-6

Chlamydiae (Chapter 13)
Actinobacteria (Chapter 14)

The two phyla, Bacteroidetes plus Fusobacteria, are grouped in the same chapter here, because they were both formerly assigned to the same (now obsolete) class, Saprospirae, the gliding, fermenting bacteria [50]. Fermentation is a metabolic process wherein organic molecules yield energy. Bacterial gliding is a process in which cells on a flat surface move through excreted slime (polysaccharides). The process of gliding is different from the process of swimming, and does not require flagella. Aside from organisms in Class Bacteroidetes and Class Fusobacteria (Chapter 10), gliding is also seen in some members of Class Gamma Proteobacteria (Chapter 7), particularly Class Pseudomonadales.

Bacteroidetes
  Bacteroidales
    Bacteroidaceae
      *Bacteroides
    Porphyromonadaceae
      *Porphyromonas
    Prevotellaceae
      *Prevotella
  Flavobacteria
    Flavobacteriales
      Flavobacteriaceae
        *Elizabethkingia

Class Bacteroidetes has two major subdivisions, Class Bacteroidales and Class Flavobacteria. The two divisions lack morphologic resemblance to one another and are physiologically distinct (Class Bacteroidales is anaerobic, Class Flavobacteria is aerobic). Nonetheless, their genotypes are similar, and some species from either group produce sphingolipids, a chemical that is seldom encountered in other eubacterial classes [21].

Members of Genus Bacteroides are mostly non-pathogenic commensals, living in the human gastrointestinal tract; some are opportunistic pathogens. Bacteroides organisms account for much of the material that composes feces. The high concentration of Bacteroides species in fecal matter is a reminder that not all intestinal bacteria belong to Class Enterobacteriaceae (Gamma Proteobacteria, Chapter 7).

Peritonitis may occur when Bacteroides species leak into the normally sterile peritoneal cavity. *Bacteroides fragilis* causes the vast majority of peritonitis cases in humans.

Members of Genus Porphyromonas live as commensals or as opportunistic pathogens in the human oral cavity. *Porphyromonas gingivalis* causes a clinically aggressive gingivitis that can lead to acute necrotizing ulcerative gingivitis (ANUG) or extend to the tissues of the mouth and face, a condition known as noma or cancrum oris.

Genus Prevotella, like Genus Porphyromonas, contains oral inhabitants that can produce plaque, halitosis, and periodontal disease. *Prevotella dentalis* produces so-called bite infections, wherein oral bacteria are inoculated, by a bite or abrasion, into adjacent tissues, producing abscesses, wound infections, or bacteremia. *Prevotella dentalis* bacteremia can lead to disseminated infections. Readers should remember that *Eikenella corrodens* (Class Beta Proteobacteria, Chapter 6) is another oral "bite" organism, that can produce a bacteremia if mechanically forced into the bloodstream.

Members of Genus Elizabethkingia are opportunistic pathogens that grow as surface biofilms. Like other organisms that grow on surfaces (see Class Pseudomonadales, Gamma Proteobacteria, Chapter 7), Elizabethkingia species cause nosocomial infections, including bacteremia associated with indwelling intravascular devices, and pneumonia associated with ventilatory support [51]. *Elizabethkingia meningoseptica*, formerly *Chryseobacterium meningosepticum*, formerly *Flavobacterium meningosepticum*, has been associated with meningitis outbreaks in neonatal intensive care units. It can also cause soft tissue infections in immune-competent individuals.

Fusobacteria
  Fusobacteriales
    Fusobacteriaceae
      *Fusobacterium
      *Streptobacillus

Members of Genus Fusobacterium are active pathogens that can colonize the oropharynx [52]. *Fusobacterium necrophorum* causes a significant percentage of acute and recurring sore throats in humans. The other significant causes of sore throats are *Streptococcus* species (Class Bacilli, Chapter 12) and viruses (e.g. Influenza virus, Rhinovirus, Coxsackie virus). *Fusobacterium necrophorum* may also cause tonsillar abscesses, and bacteremia, and is the agent responsible for Lemierre's disease (bacteraemia with disseminated abscesses, following a sore throat).

Genus Streptobacillus contains the pathogen *Streptobacillus moniliformis*. Rat-bite fever, also known as Haverhill fever, is caused by either *Streptobacillus moniliformus* or by *Spirillum minus* (Class Spirochaetes, Chapter 9) [49]. Regardless of the causative organism, or the phylogenetic classes to which the organisms are assigned, the clinical symptoms are somewhat similar, with fever and generalized complaints followed by a rash and arthritis. Deaths sometimes occur if endocarditis develops. The disease is transmitted by rodents, either by bite, close contact, or through contaminated water or food.

Infectious species:

Bacteroides fragilis (peritonitis)
Porphyromonas gingivalis (gingivitis, sometimes leading to acute necrotizing ulcerative gingivitis or ANUG, and noma, or cancrum oris)

Prevotella dentalis (abscesses, bacteraemia, wound infection, bite infections, genital tract infections, and periodontitis)
Prevotella species (supragingival plaque and halitosis in children)
Elizabethkingia meningoseptica, formerly *Chryseobacterium meningosepticum*, formerly *Flavobacterium meningosepticum* (nosocomial infections, including pneumonia, meningitis, and sepsis; cellulitis)
Fusobacterium necrophorum (Lemierre's syndrome, sore throat, peritonsillar abscess)
Streptobacillus moniliformis (rat-bite fever, Haverhill fever)

Chapter 11

Mollicutes

"The old idea that wall-less bacteria, mycoplasmas, are phylogenetically remote from other (eu)bacteria is incorrect; the true mycoplasmas are merely 'degenerate' clostridia."

Carl Woese [21]

Bacteria
  Proteobacteria
    Alpha Proteobacteria (Chapter 5)
    Beta Proteobacteria (Chapter 6)
    Gamma Proteobacteria (Chapter 7)
    Epsilon Proteobacteria (Chapter 8)
  Spirochaetes (Chapter 9)
  Bacteroidetes (Chapter 10)
  Fusobacteria (Chapter 10)
  Firmicutes (low G+C Gram+)
    Bacilli (Chapter 12)
    Clostridia (Chapter 12)
    Mollicutes (Chapter 11)
      Anaeroplasmatales
        *Erysipelothrix
      Mycoplasmataceae
        *Mycoplasma
        *Ureaplasma
  Chlamydiae (Chapter 13)
  Actinobacteria (Chapter 14)

Class Mollicutes is essentially synonymous with several other classes that appear in the literature: Class Aphragmabacteria, Class Tenericutes, and Class Mycoplasmas. Mollicutes (Latin "mollis," soft and "cutis," skin) are intracellular parasitic organisms that lack cell walls. Like many other parasitic organisms, they have jettisoned many of their inherited functionalities, preferring to live on the largesse of their hosts. Aside from shedding their

J.J. Berman: Taxonomic Guide to Infectious Diseases. DOI: http://dx.doi.org/10.1016/B978-0-12-415895-5.00011-8

cell walls (and the genes coding for cell walls), mollicutes have shrunk to about 0.2 microns; and they have a remarkably small genome (under 1000 kilobases).

The mollicutes, with no cell wall to absorb the Gram stain, are technically not Gram-positive. Nonetheless, the mollicutes are usually counted among the Gram-positive organisms, as they lack the outer membrane that is characteristic of Gram-negative organisms. Sequence similarities for ribosomal RNA and a low G+C ratio (about 35%), suggest that the mollicutes are close relatives of Class Bacilli and Class Clostridia (Chapter 12). For this reason, Class Mollicutes is often included, along with Class Bacilli and Class Clostridia, as a subclass of Class Firmicutes, the Gram-positive low G+C bacteria.

The mollicutes have exempted themselves from the strict, universal code that controls the translation of RNA into protein. Virtually every organism on earth uses the triplet codon UGA to code for "stop," thus signaling an interruption in RNA translation. Mollicutes are the exception, using UGA as a codon for tryptophan. It has been suggested that the low G+C content of mollicutes encourages the evolution of synonymous codons containing the overly abundant A (Adenine) or U (Uracil). In cells other than mollicutes, there are three synonymous "stop" codons: UAA, UAG, and UGA; while there is only one tryptophan codon: UGG. Mollicutes supplement UGG, the normal codon for tryptophan, with a codon that contains adenine and uracil, UGA.

Though it is relatively easy to detect the presence of mollicute species in human tissues, using PCR (polymerase chain reaction) techniques, these organisms cannot be cultured with any regular success. Consequently, medical scientists have implicated numerous members of Class Mollicutes as human pathogens, without fulfilling all of the rigorous studies that fully establish disease causation (i.e., they have not produced disease in humans or animals by inoculating the cultured organisms). In this chapter, we cover some of the less controversial mollicute pathogens. There are three genera of Class Mollicutes that cause diseases in humans: Erysipelothrix, Mycoplasma, and Ureoplasma.

Mollicutes
  Anaeroplasmatales
    *Erysipelothrix

Genus Erysipelothrix contains one infectious species; *Erysipelothrix rhusiopathiae*, the cause of erysipeloid, a type of cellulitis (subcutaneous infection). Erysipeloid is usually a mild condition, typically occurring on the hands of workers who are exposed to the bacteria when they handle infected fish or meat. Students should not confuse erysipeloid with the similar-sounding disease, erysipelas. Both erysipeloid and erysipelas are types of cellulitis. Erysipelas is caused by members of Genus Streptococcus (Class Bacilli,

Chapter 12) and is a more common and, potentially, more serious disease than erysipeloid. Two additional similar-sounding skin conditions are erythrasma, characterized by brown scaly skin patches; caused by *Corynebacterium minutissimum* (Class Actinobacteria, Chapter 14), and erythema infectiosum, caused by Parvovirus B19 (Chapter 40). All four skin conditions are associated with reddened skin, and all four diseases take their root from the Greek, "erusi," meaning red.

Mollicutes
  Mycoplasmataceae
    *Mycoplasma
    *Ureaplasma

Genus Mycoplasma contains two accepted human pathogens: *Mycoplasma pneumoniae* and *Mycoplasma genitalium*. *Mycoplasma pneumoniae*, as its name suggests, causes pneumonia. It is the most common cause of pneumonia in young adults. The pneumonia produced tends to be somewhat mild and chronic (so-called walking pneumonia), unlike the acute and fulminant pneumonias produced by other bacteria.

Mycoplasma genitalium is a common cause of sexually transmitted urethritis.

Mycoplasma hominis inhabits the human genital tract and is a suspected cause of some cases of pelvic inflammatory disease in women.

Mycoplasma fermentans, *Mycoplasma pirum*, and *Mycoplasma penetrans*, are additional Mycoplasma species that are being studied as potential human pathogens.

Genus Ureoplasma contains two putative infectious species: *Ureoplasma urealyticum* and the closely related *Ureoplasma parvum*. They are found in the genital tracts of a very high percentage of sexually active, healthy humans (about 70%); thus, their role as pathogens in genital and perinatal diseases is somewhat controversial. It is suspected that Ureaplasma species account for some cases of urethritis.

Infectious species:

Erysipelothrix rhusiopathiae (erysipeloid, a cellulitis)
Mycoplasma genitalium (urethritis)
Mycoplasma pneumoniae (mycoplasma pneumonia)
Ureaplasma urealyticum (urethritis)

Chapter 12

Class Bacilli Plus Class Clostridia

"We need above all to know about changes; no one wants or needs to be reminded 16 hours a day that his shoes are on."

David Hubel

Bacteria
  Proteobacteria
    Alpha Proteobacteria (Chapter 5)
    Beta Proteobacteria (Chapter 6)
    Gamma Proteobacteria (Chapter 7)
    Epsilon Proteobacteria (Chapter 8)
  Spirochaetes (Chapter 9)
  Bacteroidetes (Chapter 10)
  Fusobacteria (Chapter 10)
  Firmicutes (low G+C Gram+ bacteria)
    Mollicutes (Chapter 11)
    Bacilli (Chapter 12)
      Bacillales (catalase positive)
        Listeriaceae
          *Listeria
        Staphylococcaceae
          *Staphylococcus
        Bacillaceae
          *Bacillus
      Lactobacillales (catalase negative)
        Enterococcaceae
          *Enterococcus
        Streptococcaceae
          *Streptococcus
    Clostridia (Chapter 12)
      Clostridiales

J.J. Berman: Taxonomic Guide to Infectious Diseases. DOI: http://dx.doi.org/10.1016/B978-0-12-415895-5.00012-X

Clostridiaceae
*Clostridium
*Peptostreptococcus
Veillonellaceae
*Veillonella
Chlamydiae (Chapter 13)
Actinobacteria (Chapter 14)

Class Bacilli and Class Clostridia constitute a group of bacteria that, together, are sometimes called Class Firmicutes. In this book, the class name "Firmicutes" is abandoned because it has been used, at various times, to include Class Mollicutes (Chapter 11). Still, Class Bacilli and Class Clostridia are sister classes and share a number of important phylogenetic properties and are best discussed together.

Class Clostridia and Class Bacilli are characterized by Gram-positive species that have a low G+C ratio (a feature that distinguishes this group from Class Actinobacteria, Chapter 14, whose members are Gram-positive, with a high G+C ratio). Bacteria in Class Clostridia and Class Bacilli have a propensity for synthesizing biologically active chemicals, accounting for several of the most potent toxins in biology (e.g., botulinum toxin, tetanospasmin). Members of Class Bacilli and Class Clostridia tend to be short rods (bacilli) or round (cocci), anaerobic, and capable of forming endospores. Endospore formation, though not present in all members of Class Bacilli and Class Clostridia, is never seen outside these classes.

Bacterial endospores, often referred to by the shortened form, "spores," are fundamentally different from the spores produced by members of Class Protoctista (i.e., endospores are not the equivalent of the germinative cell of a multi-stage life cycle). Bacterial endospores are simplified forms of the bacteria, consisting of the DNA genome, some small amount of cytoplasm, and a specialized coating that confers resistance to heat, radiation, and other harsh external conditions. Endospores are virtually immortal, and can be re-activated, under favorable growth conditions, after lying dormant for hundreds or perhaps millions of years [53]. Due to their ability to grow under anaerobic conditions, and to lay dormant for long periods, it can be nearly impossible to prevent infections caused by pathogenic species of Class Bacilli and Class Clostridia.

Bacilli
Bacillales (catalase positive)
Listeriaceae
*Listeria
Staphylococcaceae
*Staphylococcus
Bacillaceae
*Bacillus
Lactobacillales (catalase negative)
Enterococcaceae

*Enterococcus
Streptococcaceae
*Streptococcus

The human pathogens in Class Bacilli are split into two groups: Class Bacillales, the catalase positive genera; and Class Lactobacillales, the catalase-negative genera.

Bacilli
Bacillales (catalase positive)
Listeriaceae
*Listeria

Listeria monocytogenes is the organism that causes listeriosis. Though *Listeria monocytogenes* is widely known as a cause of food-borne outbreaks, readers should understand that it is an opportunistic infection that rarely causes disease when it infects healthy adults. Disease, when it occurs, often manifests as sepsis or meningitis, and can be fulminant, with a high mortality (25%). Listeria is one of the few bacterial organism that produce meningoencephalitis. Other bacterial causes of meningitis include *Neisseria meningitidis* (Class Beta Proteobacteria, Chapter 6), and *Elizabethkingia meningoseptica* (Class Bacteroidete, Chapter 10). Listeriosis should not be confused with the similar-sounding disease, leptospirosis (Class Spirochaetae, Chapter 9).

Another species of Genus Listeria is *Listeria invanovii*, a pathogen for non-human animals, particularly ruminants. Like *Listeria monocytogenes*, it is opportunistic, and is a rare cause of disease in immune-compromised individuals [54]. It can produce an enteritis with bacteremia.

Bacilli
Bacillales (catalase positive)
Staphylococcaceae
*Staphylococcus

Members of Genus Staphylococcus are round cocci (from the Greek, "kokkos" meaning berry-like) despite their inclusion in Class Bacilli (from the Latin, "bacillum," meaning small rod). The two most prevalent pathogenic species are *Staphylococcus aureus* and *Staphylococcus epidermidis*.

Staphylococcus aureus lives as a non-pathogenic commensal on the skin and on the nasal mucosa of a significant portion of the human population (about 20%). It can cause skin disease (acne, skin abscesses, cellulitis, staphylococcal scalded skin syndrome) and is a common source of wound infections. If sepsis occurs, it can produce multi-organ disease, and is a cause of toxic shock syndrome in women. It is a common cause of nosocomial (hospital-acquired) infections, and strains of *Staphylococcus aureus* have emerged that are resistant to standard antibiotic treatment. The bacteria produces a toxin that causes enteritis when ingested in food that has been colonized by the bacteria.

Staphylococcus epidermidis is a commensal that lives on human skin. It is non-pathogenic in most circumstances. Chronically ill patients with indwelling catheters are prone to urinary tract infections caused by *Staphylococcus epidermidis*. This organism can grow as a biofilm, enhancing its ability to glide over surfaces (such as catheters). Nosocomial opportunistic organisms, that glide into tissues via an invasive instrumented portal (e.g., catheter, intravenous line, pulmonary assistance tubing) were described earlier with Class Pseudomonadales (Class Gamma Proteobacteria, Chapter 7) and again with the gliding bacteria (Class Bacteroidetes and Class Fusobacteria, Chapter 10).

Bacilli
  Bacillales (catalase positive)
    Bacillaceae
      *Bacillus

Bacillus cereus causes food-borne enteritis. *Bacillus cereus* endospores can survive conditions that would kill other bacterial forms. Consequently, food poisoning due to *Bacillus cereus* occurs under similar conditions as food poisoning due to *Clostridium perfringens* (see below) or *Staphylococcus aureus* (see above). Depending on the strain of *Bacillus cereus* and the conditions of its growth, an enterotoxin may accumulate in contaminated food. If preformed enterotoxin is present in contaminated food, emesis often results, within a few hours, and the condition may simulate *Staphylococcus aureus* food poisoning (see above). If no enterotoxin is present in food contaminated with *Bacillus cereus*, diarrhea usually begins after about 10 hours, and the condition may simulate infections with *Clostridium perfringens*.

Bacillus anthracis is the cause of anthrax, an acute disease that is often fatal if not treated quickly and aggressively. The disease is transmitted by endospores (not by active bacteria) that are, in most cases, spread by infected animals. Animals become infected by grazing on plants and soil containing long-dormant spores, or by eating an actively infected animal. Humans become infected by inhaling endospores emanating from the carcass of a dead infected animal (leading to pulmonary anthrax), by eating undercooked infected animals (leading to enteric anthrax), or by handling infected animals, with spores entering the skin through abrasions (leading to cutaneous anthrax). Anthrax has been weaponized by various governments over the decades, but its long dormancy and the difficulty containing spores within a specified target location, have made this weapon a double-edged sword [55].

Bacilli
  Lactobacillales (catalase negative)
    Enterococcaceae
      *Enterococcus

Species of Genus Enterococcus are normal inhabitants of the gastrointestinal tract. Disease most often occurs in a hospital setting, in weakened patients who have had surgery, or who have indwelling devices (e.g., urinary catheters). Urinary tract infections or bacteremia and its sequelae (e.g., endocarditis, meningitis) have been associated with *Enterococcus faecalis* and *Enterococcus faecium* species.

Bacilli
 Lactobacillales (catalase negative)
  Streptococcaceae
   *Streptococcus

Genus Streptococcus contains numerous pathogenic and non-pathogenic species (too many to describe here).

Streptococcus pneumoniae, as its name suggests, causes pneumonia. It may also cause disease via nasopharyngeal spread (e.g., otitis media, sinusitis, meningitis).

Streptococcus pyogenes is a very common cause of human infection, producing pharyngitis (strep throat), and skin infections (streptococcal impetigo). Toxin-producing strains can produce scarlet fever (systemic symptoms plus rash). Bacteremia may lead to toxic shock syndrome. Following infection with *Streptococcus pyogenes*, an immune response cross-reacting between bacterial antigens and normal host proteins (e.g., muscle proteins, glomerular basement membranes), may lead to rheumatic fever or glomerulonephritis.

Clostridia
 Clostridiales
  Clostridiaceae
   *Clostridium
   *Peptostreptococcus

Class Clostridia is comprised of obligate anaerobic organisms (unlike Class Bacilli, which includes facultative anaerobes). Class Clostridia has two genera that contain human pathogens: Clostridium and Peptostreptococcus.

Genus Clostridium: contains four species that commonly produce disease in humans: *Clostridium botulinum*, *Clostridium difficile*, *Clostridium perfringens*, and *Clostridium tetani*.

Clostridium botulinum produces botulinum toxin, one of the most powerful poisons in existence. Disease in adults is caused by ingesting the pre-formed toxin produced by bacteria growing in contaminated food. Because all members of Class Clostridia are obligate anaerobes, contamination occurs in foods stored in anaerobic conditions (e.g., cans), without first killing all the bacteria and spores. When adults ingest *Clostridium bacilli*, the bacteria are usually killed by competing organisms in the intestinal tract. The ingested toxin causes the disease known as botulism. In infants, ingested bacteria may survive in the intestinal tract, actively producing toxin. Spores of *Clostridium botulinum* in honey have been known to produce active *Clostridium botulinum* infection in children.

Clostridium difficile inhabits human intestine and is non-pathogenic under normal circumstances. After long-term antibiotic use, when many of the normal gut bacteria are reduced in number, an overgrowth of *Clostridium difficile* may cause severe gastrointestinal disease (so-called pseudomembranous colitis). Colon ulcerations, with an overlying pseudomembrane composed of necrotic mucosal cells admixed with inflammatory cells, is the hallmark of this disease.

Clostridium perfringens is a ubiquitous organism that is sometimes found in the human gastrointestinal tract, without causing disease. It is a common cause of food poisoning. When contaminated food is ingested, diarrhea often follows, in about 10 hours, in susceptible individuals (some individuals are resistant to enteric disease). *Clostridium perfringens* is a common infection in necrotic tissue, due to the anoxic conditions therein. The organism causes so-called gas gangrene (tissue necrosis accompanied by the liberation of bacterial-produced gas). *Clostridium perfringens* also causes emphysematous gangrenous cholecystitis, a condition occurring with gallbladder necrosis, in which the necrotic gallbladder tissue is infiltrated by gas, produced by the organism. The ability of *Clostridium perfringens* to produce gas has a beneficial purpose: as a leavening agent for baked goods.

Clostridium tetani is the cause of tetanus. Spores live in soil, and human infection usually follows the mechanical introduction of soil-borne spores into a wound. The organism produces a potent neurotoxin that manifests clinically as muscle rigidity: risus sardonicus (rigid smile), trismus (also known as lock-jaw, rigid jaw), and opisthotonus (rigid, arched back).

Though only members of Class Bacilli and Class Clostridia have the ability to form endospores, not all members of these two classes are spore-forming. Genus Peptostreptococcus is the exception. Species of genus Peptostreptococcus are found as commensals in virtually every type of mucosa that lines humans. They have pathogenic potential when they are traumatically introduced deep into tissues, or when the host becomes weakened from concurrent chronic infections, or when the host becomes immune-deficient. Under these circumstances, they can produce sepsis, with abscesses occurring in multiple organs.

Clostridia
  Veillonellaceae
    *Veillonella

Species in Genus Veillonella are normal inhabitants of the GI tract of humans and other mammals. As with infections from Peptostreptococcus species, Veillonella species may cause sepsis, with multi-organ disease in predisposed individuals. Veillonella infections are rare.

Infectious species:

Listeria ivanovii (in immune-deficient hosts)
Listeria monocytogenes (listeriosis)
Staphylococcus aureus (staphylococcal scalded skin syndrome, toxic shock syndrome, acne, skin abscesses, cellulitis, sepsis, food poisoning)
Staphylococcus epidermidis (catheter-transmitted urinary tract infections, commonly found in acne infections)
Bacillus cereus (fried rice syndrome, small fraction of food-borne illnesses)
Bacillus anthracis (anthrax)
Enterococcus species (bacteremia, diverticulitis, endocarditis, meningitis, urinary tract infections)
Streptococcus pneumoniae (pneumococcal pneumonia, childhood meningitis)
Streptococcus agalactiae (bacterial septicemia of the newborn) [56]
Streptococcus iniae (bacteremic cellulitis, sepsis, in immune-deficient individuals)
Streptococcus pyogenes (scarlet fever, erysipelas, rheumatic fever, streptococcal pharyngitis, post-streptococcal glomerulonephritis)
Streptococcus suis (multi-organ infections)
Clostridium botulinum (botulism)
Clostridium difficile (pseudomembranous colitis)
Clostridium perfringens (gas gangrene, clostridial necrotizing enteritis, food poisoning, emphysematous cholecystitis)
Clostridium tetani (tetanus)
Peptostreptococcus species, formerly Peptococcus species, including *Peptostreptococcus magnus* (septicemia, organ abscesses, cellulitis, particularly in immune-compromised patients)

Chapter 13

Chlamydiae

"Simplicity, carried to the extreme, becomes elegance."

Jon Franklin

- Bacteria
  - Proteobacteria
    - Alpha Proteobacteria (Chapter 5)
    - Beta Proteobacteria (Chapter 6)
    - Gamma Proteobacteria (Chapter 7)
    - Epsilon Proteobacteria (Chapter 8)
  - Spirochaetes (Chapter 9)
  - Bacteroidetes (Chapter 10)
  - Fusobacteria (Chapter 10)
  - Firmicutes
    - Bacilli (Chapter 12)
    - Clostridia (Chapter 12)
    - Mollicutes (Chapter 11)
  - Chlamydiae (Chapter 13)
    - Chlamydiae
      - Chlamydiales
        - Chlamydiaceae
          - *Chlamydophila
          - *Chlamydia
  - Actinobacteria (Chapter 14)

All members of Class Chlamydiae are obligate intracellular pathogens (like the Rickettsia, members of Alpha Proteobacteria covered in Chapter 5). All Chlamydiaceae are Gram-negative, and they all express the same lipopolysaccharide epitope, that has only been observed in Class Chlamydiaceae. Members of Class Chlamydiaceae are extremely small, less than 1 micron in size. These organisms grow exclusively within eukaryotic cells. Two genera of Chlamydiaceae contain human pathogens: Chlamydia and Chlamydophila. Genus Chlamydia and Genus Chlamydophila are closely related and, prior to

J.J. Berman: Taxonomic Guide to Infectious Diseases. DOI: http://dx.doi.org/10.1016/B978-0-12-415895-5.00013-1

1999, all of the group species were assigned to Genus Chlamydia. Molecular studies indicated that these two genera can be cleanly distinguished from one another, based on genome size, DNA re-association, and sequence dissimilarities.

Most obligate intracellular infectious agents require a vector to transfer themselves from one host to another. For example, Rickettsia (Chapter 5), another obligate intracellular bacteria, are transmitted via fleas, ticks, or lice. Malaria and babesiosis, caused by obligate intracellular members of Class Apicomplexa (Chapter 19), are transmitted by arthropods. Surprisingly, the human pathogens in Class Chlamydiaceae manage to move from one host to another, without the aid of a vector. How do they do it? These tiny organisms create even smaller, infective forms, known as elementary bodies. Elementary bodies have a rigid outer membrane and are resistant to environmental conditions outside their hosts. They travel in expelled droplets, in the case of a pneumonic infection, or in secretions, in the case of a venereal infection or an eye infection. The elementary bodies attach to host cell membranes, and are internalized within host cell endosomes. Elementary bodies transform into reticulate bodies, the metabolically active form of the organism. Chlamydia and Chlamydophila organisms inhibit the fusion of host endosomes with host lysosomes, and thus escape the normal cellular mechanism by which phagocytosed bacteria are killed by eukaryotes. This bacterial survival trick is similar to that employed by *Coxiella burnetii* (Chapter 7), another obligate intracellular bacteria that is resistant to degradation by host cells. When the endosome is filled with organisms of Class Chlamydiaceae, the enlarged endosome becomes a cytoplasmic inclusion body, visible under the light microscope. Active infections are characterized by eukaryotic cell lysis. Most cases of infection with Chlamydia or Chlamydophila are asymptomatic, indicating that cytopathic effects are often minimal. Immune deficiency exacerbates the clinical virulence of infections caused by pathogenic members of Class Chlamydiae.

Chlamydiae
  Chlamydiae
    Chlamydiales
      Chlamydiaceae
        *Chlamydophila
        *Chlamydia

In terms of documented infections and disease, the most clinically important species in Class Chlamydiaceae is *Chlamydia trachomatis*. This species can be divided into several biological types, and the biological types can be subdivided into distinct serologic variants (serovars). The different diseases caused by *Chlamydia trachomatis* are each associated with their own variants of the species. Estimates would suggest that worldwide, more than half a billion people are infected with one or another subtype of *Chlamydia trachomatis*.

Infection by *Chlamydia trachomatis* is the second most common sexually transmitted disease, with about 4 million new cases occurring annually in North America [32]. The disease name for this infection is somewhat confusingly called "Chlamydia," perhaps the only disease that takes the name of a genus, without modification. The proper disease name "chlamydiosis" is reserved for infections by another species, *Chlamydophila psittaci* (see below). This is an etymologic disaster, as *Chlamydophila psittaci* infection is known by most clinicians as psittacosis, and would more accurately be called chlamydophilosis, in any event. In men and women, Chlamydia can produce urethritis and rectal inflammation. Chlamydia can also produce prostatitis in men. In women, infections that ascend the genital tract can yield endometritis, salpingitis, and pelvic inflammatory disease. Infants born to infected mothers may develop inclusion conjunctivitis (named for the cytoplasmic inclusion bodies produced by chlamydial organisms), and chlamydial pneumonia.

The high infection rate in the population is made possible, in part, by the high prevalence of carriers: about one third of infected men and women have no clinical symptoms. In addition, infection does not confer immunity, and re-infections are common.

According to the World Health Organization, there are about 37 million blind persons, worldwide. Trachoma, caused by *Chlamydia trachomatis*, is the number one infectious cause of blindness and accounts for about 4% of these cases. The second most common infectious cause of blindness worldwide is *Onchocerca volvulus* (Class Nematoda, Chapter 27), accounting for about 1% of cases [28]. Trachoma is spread by direct or indirect contact with eye secretions. Infection causes intense inflammation of the conjunctiva.

Readers should not confuse trachoma with inclusion conjunctivitis, as each disease is caused by distinct variants of the same species (*Chlamydia trachomatis*). Trachoma is contracted by exposure to eye secretions from people with trachoma. Inclusion conjunctivitis is caused by ocular exposure to secretions from the sexually transmitted infection.

Chlamydia trachomatis may also cause lymphogranuloma venereum, a disease that usually presents as swollen lymph nodes in the groin. The lymph nodes often have draining abscesses. The disease is rare, with only a few hundred cases occurring in the United States each year. Lymphogranuloma venereum must not be confused with granuloma inguinale, also known as granuloma venereum, caused by the bacterium *Klebsiella granulomatis* (Chapter 7).

Chlamydophila psittaci is the cause of psittacosis, a disease that takes its name from birds of Class Psittaciformes (i.e., parrots) the organism's animal reservoir. Chlamydophila infects the lungs of birds. The birds pass the infection to humans through droplet secretions.

Chlamydophila pneumoniae is a significant cause of pneumonia in the United States. The disease is often mild. In addition, *Chlamydophila pneumoniae* has been associated with coronary artery disease and stroke [6]. The organism can infect the endothelial cells of coronary arteries, and has been found in atherosclerotic plaques. At this time, a causative role in atherogenesis has not been demonstrated.

Infectious species:

Chlamydophila pneumoniae, formerly TWAR serovar, TWAR agent, or *Chlamydia pneumoniae* (pneumonia)
Chlamydophila psittaci (psittacosis)
Chlamydia trachomatis (trachoma, genital infection)

Chapter 14

Actinobacteria

"It is once again the vexing problem of identity within variety; without a solution to this disturbing problem there can be no system, no classification."

Roman Jakobson

Bacteria
  Proteobacteria
    Alpha Proteobacteria (Chapter 5)
    Beta Proteobacteria (Chapter 6)
    Gamma Proteobacteria (Chapter 7)
    Epsilon Proteobacteria (Chapter 8)
  Spirochaetes (Chapter 9)
  Bacteroidetes (Chapter 10)
  Fusobacteria (Chapter 10)
  Firmicutes, low G+C Gram+
    Bacilli (Chapter 12)
    Clostridia (Chapter 12)
    Mollicutes (Chapter 11)
  Chlamydiae (Chapter 13)
  Actinobacteria, high G+C Gram+ (Chapter 14)
    Actinomycetales
      Actinomycetaceae
        *Actinomyces
        *Arcanobacterium
      Corynebacterineae
        Corynebacteriaceae
          *Corynebacterium
        Dermatophilaceae
          *Dermatophilus
        Mycobacteriaceae
          *Mycobacterium

J.J. Berman: Taxonomic Guide to Infectious Diseases. DOI: http://dx.doi.org/10.1016/B978-0-12-415895-5.00014-3

Nocardiaceae
*Nocardia
*Rhodococcus
Cellulomonadaceae
*Tropheryma
Propionibacteriaceae
*Propionibacterium
Streptosporangineae
Thermomonosporaceae
*Actinomadura
Bifidobacteriales
Bifidobacteriaceae
*Gardnerella

There are two large classes within the Gram-positive group (not all of which are actually Gram-positive). These are the Firmicutes (containing Class Bacilli (Chapter 12), Class Clostridia (Chapter 12), and Class Mollicutes (Chapter 11)) and the Actinobacteria (Chapter 14). The Firmicutes are characterized by low G+C DNA. The Actinobacteria are characterized by high G+C DNA.

Members of Class Actinobacteria tend to be filamentous, and this morphologic feature has led to great confusion. In the past, these filamentous bacteria were mistaken for fungal hyphae, and many of the diseases caused by members of Class Actinobacteria were mistakenly assigned fungal names (e.g., actinomycosis, mycetoma, maduromycosis).

Molecular analysis of the Actinobacteria indicates that they are eubacteria that share a high degree of sequence similarity among the subclasses [21]. High sequence similarity among sister subclasses is generally interpreted to mean that the classes are young (i.e., descended from a common ancestor relatively recently, before their genomes had an opportunity to diverge).

Actinobacteria
Actinomycetales
Actinomycetaceae
*Actinomyces
*Arcanobacterium

Actinomyces is the cause of so-called actinomycosis (Greek "actino," ray and "myco," fungus), a suppurative condition arising in the oral cavity and nasopharynx, characterized by acute and chronic inflammation and the discharge of so-called sulfur granules. The disease is caused by any of several species of Genus Actinomyces (e.g., *Actinomyces israelii* or *Actinomyces gerencseriae*), as well as closely related species (e.g., *Propionibacterium propionicus*). Sulfur granules are yellow flecks found mixed with inflammation and extruded in exudate, consisting of numerous bacterial filaments, often radiating from a core (hence the prefix "actino") and resembling fungal hyphae (hence the suffix "mycosis"). Actinomyces species inhabit normal mouths, and sulfur granules are frequently found in the tonsillar crypts of healthy individuals.

Several species closely related to Genus Actinomyces can produce an allergic disease of the lung after chronic inhalation (e.g. *Micropolyspora faeni* and less commonly *Saccharopolyspora rectivirgula*). This disease, which is not a true infection, is known by a number of different names, including farmer's lung.

Genus Arcanobacterium contains *Arcanobacterium haemolyticum*, formerly assigned to a different genus, as *Corynebacterium haemolyticum*. *Arcanobacterium haemolyticum* is a cause of pharyngitis.

- Actinobacteria
  - Actinomycetale
    - Corynebacterinae
      - Corynebacteriaceae
        - *Corynebacterium
      - Dermatophilaceae
        - *Dermatophilus
      - Mycobacteriaceae
        - *Mycobacterium
      - Nocardiaceae
        - *Nocardia
        - *Rhodococcus

Members of Class Corynebacterineae share a distinctive and complex cell wall composition. Most species contain mycolic acids in their cells walls, linked to peptidoglycans. Class Corynebacterineae, like all the members of Class Actinobacteria, are Gram-positive. Class Corynebacterineae (particularly Genus Mycobacteria and Genus Nocardia) resist decolorization with acid. The term "acid fast" refers to this property, in which the Gram stain is "fastened" to the cell wall. Mycolic acid contributes to "acid-fastness." Pathologists employ so-called acid-fast stains (technically, the organism is acid-fast, not the stains) to identify pathogenic organisms in this group.

Class Corynebacterium, a subclass of Class Corynebacterineae, contains several pathogenic species.

Corynebacterium minutissimum causes erythrasma, a skin rash. *Corynebacterium jeikeium*, which is normally confined to skin, tracks into blood via indwelling devices, causing opportunistic nosocomial infections (e.g., sepsis, endocarditis) in immune-compromised patients, particularly those who have received bone marrow transplants [57].

Corynebacterium diphtheriae is the cause of diphtheria, a disease that occurs where vaccination is underutilized. The disease usually presents as a sore throat, covered with a characteristic membrane. The term diphtheria has its root in the Greek work, "diphthera" meaning two leather scrolls, referring to the thick, bilateral membrane. The most aggressive strains of the organism produce a toxin, encoded by a phage (bacterial virus) in the genome. The toxin contributes to tissue necrosis. Some cases of diphtheria involve the skin.

Readers should be aware of the highly confusing term, "diphtheroid," commonly applied to all the non-pathogenic species within Genus Corynebacterium. As non-pathogens, the diphtheroids do not cause diphtheria. Diphtheria is caused *Corynebacterium diphtheriae* (i.e., a non-diphtheroid).

Species in Genus Dermatophilus produce skin lesions in a variety of animals, including humans [58]. Pitted keratolysis is a skin condition characterized by the appearance of discoloration on the palms or the plantar surfaces, with superficial craters. It may arise from various species of Genus Dermatophilus, *Dermatophilus congolensis* among them [59]. The organisms are found in the environment, and human-to-human contagion does not seem to be a mode of transmission.

About 2 billion people (of the world's 7 billion population) have been infected with *Mycobacterium tuberculosis*. Most infected individuals never develop overt disease. Nonetheless, tuberculosis kills about 3 million people each year [4]. Disease is transmitted from humans who have active, untreated disease, often by aerosolized droplets. The disease usually presents as a granulomatous process in the lungs. Multi-organ involvement may occur if the lung disease is not adequately treated. Aside from *Mycobacterium tuberculosis*, several other species may cause a tuberculosis-like disease. These species include: *Mycobacterium bovis*, *Mycobacterium africanum*, *Mycobacterium canetti*, and *Mycobacterium microti*.

Mycobacterium xenopi produces a chronic pulmonary disease simulating tuberculosis. *M. xenopi* does not seem to be transmitted from person to person. The organism has been isolated from water and soil, and the presumed mode of transmission is through environmental exposure. Though the disease is currently rare in the USA, cases are not unusual in England, Europe, and Canada.

Leprosy, also known as Hansen's disease, is caused by *Mycobacterium leprae* and *Mycobacterium lepromatosis*. The disease produces granulomas in the skin, including the nerves of the skin, and the respiratory tract. Untreated leprosy is slowly progressive. Most human cases seem to be transmitted by aerosolized droplets produced by actively infected and untreated patients. It is believed that animals, particularly the armadillo, are potential reservoirs. New cases of leprosy have been dropping. In 2005, there were about 300 000 new cases reported, worldwide [60].

Mycobacterium ulcerans causes Buruli ulcer, a skin condition seen in the tropics. *Mycobacterium ulcerans* produces a toxin, mycolactone, that is responsible for most of the tissue destruction observed clinically. Other mycobacteria that produce mycolactone include *Mycobacteria liflandii*, *Mycobacteria pseudoshottsii*, and strains of *Mycobacteria marinum*. The disease does not appear to be spread from human to human; aquatic insects are the suspected vectors.

The atypical mycobacteria are a group of about 13 organisms that are found in soil and water, as pathogens in animals, or even as commensals growing in the pharynx of humans. Diseases, when they occur in otherwise healthy individuals, tend to be mild, involving skin, lungs, or lymph nodes. The atypical mycobacteria pose a serious problem in immune-compromised patients, particularly those with AIDS. Organisms in the *Mycobacterium-avium-intracellulare* complex, found throughout the environment, are a particular threat for patients with advanced AIDS.

Genus Nocardia contains dozens of species that cause disease in animals. In humans, Nocardia organisms are opportunistic pathogens affecting children, the elderly, immune-compromised individuals, and patients with a pre-existing serious disease. *Nocardia asteroides* is the most common cause of human nocardiosis. Pneumonia is a common presentation of nocardiosis. Encephalitis, endocarditis, abscess formation, and sepsis are also seen. *Nocardiosis brasiliensis* has been implicated in some cases of mycetoma. A full discussion of mycetoma is found under Genus Neotestudina (Ascomycota, Chapter 36).

Readers should not be confused by the plethora of organisms with "brasiliensis" as the species of the binomen. These include *Nocardia brasiliensis*, *Leishmania brasiliensis* (alternately spelled *Leishmania braziliensis*), *Paracoccidioides brasiliensis*, and *Borrelia brasiliensis*.

Rhodococcus equi is the cause of foal pneumonia, or rattles, in young horses. It can infect immune-compromised humans, causing a disease that closely simulates tuberculosis.

Actinobacteria
  Actinomycetales
    Cellulomonadaceae
      *Tropheryma

The only pathogenic species in Class Cellulomonadacea is *Tropheryma whipplei*, the cause of Whipple disease. As a general rule, bacteria in the human body are eaten by macrophages, wherein they are degraded. In the case of *Tropheryma whipplei*, certain susceptible individuals have a problem with degradation of the organism within macrophages. Consequently, the organisms multiply within macrophages. When organisms are released from dying macrophages, additional macrophages arrive to feed, but this only results in the local accumulation of macrophages bloated by bacteria. Whipple disease is characterized by the infiltration of organs by foamy macrophages containing *Tropheryma whipplei*. The organ most often compromised is the small intestine, where infiltration of infected macrophages in the lamina propria (the connective tissue underlying the epithelial lining of the small intestine) causes malabsorption. Whipple disease is quite rare. It occurs most often in farmers and gardeners who work with soil, in which the organism lives. As recently as 1992, the cause of Whipple disease was unknown [61].

Readers should not confuse the bacterial genus Tropheryma with the similar sounding term Taphrinomycotina, the fungal class that includes Pneumocystis (Class Ascomycota, Chapter 36).

Actinobacteria
  Actinomycetales
    Propionibacteriaceae
      *Propionibacterium

Class Propionibacteriaceae is named for a particular metabolic talent, propionic acid synthesis. Species in the class live in the intestinal tracts of various animals. In humans, species of Genus Propionibacterium are commensals that live within sweat glands. Some cases of acne and other skin conditions have been attributed to propionibacteria.

Actinobacteria
  Actinomycetales
    Streptosporangineae
      Thermomonosporaceae
        *Actinomadura

Genus Actinomadura contains species found in soil. These filamentous organisms were previously mistaken for fungal hyphae. The inflammatory skin condition, most commonly occurring on the foot, and caused by bacterial species of the Actinomadura genus, were given the fungal misnomers actinomycetoma and maduromycosis. *Actinomadura madurae*, like *Nocardiosis brasiliensis* (see above), has been implicated in some cases of mycetoma. A full discussion of mycetoma is found under Genus Neotestudina (Ascomycota, Chapter 36).

Persons who are immune-compromised may have heightened susceptibility to systemic infections with Actinomadura species [62].

Actinobacteria
  Bifidobacteriale
    Bifidobacteriaceae
      *Gardnerella

Gardnerella vaginalis is the only pathogenic species of Genus Gardnerella. It is sexually transmitted and colonizes the genital mucosa without producing active disease in most women. Overgrowth of *Gardnerella vaginalis* results in a condition known as vaginosis, producing vaginitis accompanied by a frothy discharge. Gardnerella overgrowth can be observed in routine Pap smears as tiny organisms matting the surface of flattened squamous epithelial cells (so-called "clue cells").

Infectious species:

Actinomyces gerencseriae (dental plaque)
Actinomyces israelii (actinomycosis)

Arcanobacterium haemolyticum, formerly *Corynebacterium haemolyticum* (pharyngitis)
Corynebacterium diphtheriae (diphtheria)
Corynebacterium minutissimum (erythrasma)
Corynebacterium pseudotuberculosis (ulcerative lymphangitis in horses and cattle, rarely causing lymphadenitis in humans)
Corynebacterium jeikeium (sepsis)
Dermatophilus sp. (dermatophilosis, pitted keratolysis)
Dermatophilus congolensis (dermatophilosis, mud fever, pitted keratolysis)
Mycobacterium abscessus (chronic pulmonary disease, wound infections, in immune-compromised patients)
Mycobacterium avium (persistent cough, can cause disseminated disease, including bone marrow infection, in immune-compromised individuals, collarstud abscess of neck lymph node in children)
Mycobacterium haemophilum (collarstud abscess of neck lymph node in children)
Mycobacterium intracellulare (persistent cough, can cause disseminated disease in immune-compromised individuals)
Mycobacterium scrofulaceum (cervical lymphadenitis in children)
Mycobacterium chelonae (granulomatous and acute inflammatory infections of skin and soft tissues)
Mycobacterium fortuitum (pulmonary diseases, post-surgical wound abscesses, sepsis with multi-organ involvement)
Mycobacterium kansasiim (aquarium granuloma)
Mycobacterium leprae (leprosy, Hansen disease)
Mycobacterium lepromatosis (leprosy, Hansen disease)
Mycobacterium malmoense (cervical lymphadenitis in children, pulmonary disease in adults with pre-existing lung conditions)
Mycobacterium marinum (aquarium granuloma)
Mycobacterium paratuberculosis (suspected to cause some cases of Crohn's disease)
Mycobacterium simiae (granulomatous lung disease)
Mycobacterium szulgai (tuberculosis-like pulmonary infection, disseminated disease in immune-compromised individuals)
Mycobacterium tuberculosis complex, including *Mycobacterium caprae*, *Mycobacterium tuberculosis*, *Mycobacterium africanum*, *Mycobacterium bovis*, *Mycobacterium bovis* BCG, *Mycobacterium microti*, *Mycobacterium canettii*, *Mycobacterium pinnipedii*, *Mycobacterium mungi* (tuberculosis)
Mycobacterium tuberculosis (tuberculosis)
Mycobacterium ulcerans (Buruli ulcer)
Mycobacterium xenopi (*Mycobacterium xenopi* pneumonia)
Nocardia asteroides (nocardiosis)
Nocardia brasiliensis (nocardiosis)
Nocardia caviae (nocardiosis)

Nocardia farcinica (nocardiosis)
Nocardia nova (nocardiosis)
Nocardia otitidiscaviarum (nocardiosis)
Rhodococcus equi, formerly *Corynebacterium equi*, formerly *Bacillus hoagii*, formerly *Corynebacterium purulentus*, formerly *Mycobacterium equi*, formerly *Mycobacterium restrictum*, formerly *Nocardia restricta*, formerly *Proactinomyces restrictus* (chronic pulmonary infection simulating tuberculosis)
Tropheryma whipplei, formerly *Tropheryma whippelii* (Whipple disease)
Actinomadura madurae (mycetoma, maduromycosis, madura foot)
Actinomadura pelletieri (mycetoma)
Gardnerella vaginalis, formerly *Corynebacterium vaginalis*, formerly *Haemophilus vaginalis* (vaginosis)

Part III

Eukaryotes

Chapter 15

Overview of Class Eukaryota

"The evolution of sex is the hardest problem in evolutionary biology."

John M. Smith

On the simplest level, all cellular life can be divided into two forms: the eukaryotes, which have a membrane-bound organelle, known as the nucleus, that contains the cell's genetic material; and the prokaryotic life forms (Eubacteria and the Archaeans, also known as the Archaebacteria) that have no nucleus.

Nobody knows when the first eukaryotes appeared on earth. One school of thought estimates that eukaryotes were here 2.7 billion years ago, about 1 billion years following the first appearance of prokaryotic life forms. This theory is based on finding sterane molecules in shale rocks, dating back nearly 3 billion years. Eukaryotic cells are the only known source of naturally occurring sterane molecules [63]. Other biologists tie their estimate of the beginning of eukaryotic life to the epoch in which the first eukaryotic fossil remains are found, about 1.7 billion years ago. This leaves a 1 billion year gap between scientific estimates for the origin of the eukaryotes (i.e., 1.7 to 2.7 billion years ago).

Though eukaryotes are distinguished from prokaryotes by the presence of a membrane-delimited nucleus, it would seem that the very earliest eukaryotes, and all of their living descendants, came equipped with two additional structures that separate them from prokaryotes:

1. Mitochondria. Mitochondria are membrane-delimited organelles, with their own genome, that proliferate within the eukaryotic cell. Current theory holds that mitochondria developed as an obligate intracellular endosymbiont from an ancestor of Class Rickettsia. All existing eukaryotic organisms descended from ancestors that contained mitocondria. Furthermore, all existing eukaryotic organisms, even the so-called amitochondriate classes (i.e., organisms without mitochondria), contain either fully functional or vestigial forms of mitochondria (i.e., hydrogenosomes and mitosomes) [64–67].

J.J. Berman: Taxonomic Guide to Infectious Diseases. DOI: http://dx.doi.org/10.1016/B978-0-12-415895-5.00015-5

2. Undulipodia. Prokaryotes and eukaryotes have flagella, rods that protrude from the organism; their back-and-forth motion propels cells forward through water. Aside from a superficial resemblance, the flagella of eukaryotes have no relationship to the flagella of prokaryotes. Eukaryotic flagella are orders of magnitude larger than the prokaryotic flagella, contain hundreds of proteins not present in the flagella of prokaryotes, have a completely different internal structure, anchor to a different cellular location, and do not descend phylogenetically from prokaryotic flagella [68]. Biologists provided the eukaryotic flagellum with its own name: undulipodium. Perhaps they chose a term with a few too many syllables. Most biologists continue to use the misleading term "flagellum" (plural, "flagella") to apply to prokaryotes and to eukaryotes. Regardless, every existing eukaryote descended from an organism with a undulipodium. In humans, cilia on the surface of mucosal lining cells are modified undulipodia. Every human spermatocyte has a tail with one posterior undulipodium.

As far as anyone knows, the very first eukaryote came fully equipped with a nucleus, one or more undulipodium, and one or more mitochondria. Based on similarities between the eukaryotic nucleus and archaean cells, in terms of the structure and organization of DNA, RNA, and ribosomes, it has been hypothesized that the eukaryotic nucleus was derived from an archaean organism. Conversely, similarities between mitochondria and eubacteria of Class Rickettsia suggest that the eukaryotic mitochondrium was derived from an ancestor of a modern Rickettsia. A single eukaryotic cell may contain thousands of mitochondria, indicating the huge size differential between prokaryotes and eukaryotes. Regarding the origin of the undulipodium, it is a profound mystery.

For well over a century, biologists had a very simple way of organizing the eukaryotes [12]. Basically, the one-celled eukaryotes were called protists. The multi-celled eukaryotes were assigned to the kingdom of plants or the kingdom of animals, often referred to as flora and fauna, respectively. The fungi were considered a subclass of plants. To this day, university-based mycologists (specialists in fungi) are typically employed in the Botany Department (not the Fungus Department). The protists, also known as protoctista, were considered the ancestral organisms for the multicellular organisms. Biologists applied the term "protozoa" to protist species that were the ancestral forms of animals (from "proto," the first and "zoa," animals; see Glossary item, Protozoa).

In the past decade, taxonomists have abandoned the venerable Class Protoctista. The former Class Protoctista was a grab-bag of unrelated classes that did not fit under Class Animalia, Class Fungi, or Class Plantae. Furthermore, as the phylogenetic lineage of the various eukaryotes became better understood, it became clear that the multicellular classes had much closer relationships to some of the single-celled organisms than those single-celled organisms had to other organisms in Class Protoctista. Preserving the concept of Class Protoctista was making it difficult to appreciate the true phylogenetic relationships among the eukaryotes.

Modern classifications of eukaryotic organisms simply dispense with Class Protoctista, assigning each individual class of eukaryotes to a hierarchical position determined by ancestry. A simple schema demonstrates the modern classification of eukaryotes, and also serves to organize the chapters in Part III of this book.

- Eukaryota (organisms that have nucleated cells)
  - Bikonta (2-flagella)
    - Excavata
      - Metamonada (Chapter 16)
      - Discoba
        - Euglenozoa (Chapter 17)
        - Percolozoa (Chapter 18)
    - Archaeplastida, from which Kingom Plantae derives (Chapter 24)
    - Chromalveolata (Chapters 19–21)
      - Alveolata
        - Apicomplexa (Chapter 19)
        - Ciliophora (ciliates) (Chapter 20)
      - Heterokontophyta (Chapter 21)
  - Unikonta (1-flagellum)
    - Amoebozoa (Chapter 22)
    - Opisthokonta
      - Choanozoa (Chapter 23)
      - Animalia (Chapters 25–32)
      - Fungi (Chapters 33–37)

Class Animalia (Chapters 25–32) and Class Fungi (Chapters 33–37) are much more closely related to Class Choanozoa (Chapter 23), than they are to Class Plantae (Chapter 24). Likewise, Class Choanozoa is more closely related to Class Animalia and to Class Fungi than they are to Class Metamonada (Chapter 16). If we were still using Class Protoctista, we would have assigned both Class Choanozoa and Class Metamonada to the same superclass, and we would have blurred the relationships between the subclasses. Today, the term "protist" or "protoctista" simply refers to any member of Class Eukaryota that is neither plant, animal, nor fungus.

A quick glance at the eukaryotic schema indicates that the very first division in the classification of the eukaryotes, is based on the number of flagella: Class Bikonta (from the Greek "kontos," meaning pole) consists of all organisms with two undulipodia; and Class Unikonta consists of all organisms with one undulipodia [69]. At first blush, the number of undulipodia would appear to be a poor way of distinguishing classes of organisms; it would seem likely that descendants of a bikont might evolve to lose one flagellum, or unikonts might evolve to grow a second flagellum. If this were to happen, the phylogenetic distinction between unikonts and bikonts would be lost.

As it turns out, the early division of eukaryotes into unikonts and bikonts is one of the most brilliant achievements in modern taxonomy. Whether a modern class eukaryotes has gained or lost a flagellum makes little difference. In fact, Class Fungi, a subclass of Class Unikonta, has evolved to lose its flagellum. No matter; the ancestry of Class Fungi can be traced to Class Unikonta. Three fused genes (carbamoyl phosphate synthase, dihydroorotase, aspartate carbamoyltransferase) are uniquely characteristic of Class Unikonta. Two fused genes (thymidylate synthase and dihydrofolate reductase) uniquely characterize Class Bikonta. The morphologic property dividing Class Eukaryota into unikonts and bikonts is shadowed by a genetic property that draws the equivalent taxonomic division.

The value of the eukaryotic flagellum (undulipodium) as a taxonomic divider is witnessed again in Class Opisthokonta. Class Opisthokonta is a subclass of Class Unikonta that contains Class Choanozoa, Class Animalia and Class Fungi. The opisthokonts all descend from an organism with its undulipodium extending from the rear (from the Greek "opisthios," meaning rear and "kontos" meaning pole). The rear-ended flagellum distinguishes the members of Class Opisthokonta from classes that have a flagellum extending from anterior or lateral sides.

Chapters 16 through 37 will demonstrate that the human eukaryotic pathogens can be classified in terms of their inherited class properties.

Chapter 16

Metamonada

"Our greatest responsibility is to be good ancestors."

Jonas Salk

- Eukaryota
  - Bikonta (2-flagella)
    - Excavata
      - Metamonada (Chapter 16)
        - Trichozoa
          - Parabasalidea
            - Trichomonadida
              - Trichomonadidae
                - *Trichomonas
              - Monocercomonadidae
                - *Dientamoeba
        - Fornicata
          - Diplomonadida
            - Hexamitidae
              - *Giardia
      - Discoba
        - Euglenozoa (Chapter 17)
        - Percolozoa (Chapter 18)
    - Archaeplastida (Chapter 24)
    - Chromalveolata (Chapters 19–21)
      - Alveolata
        - Apicomplexa (Chapter 19)
        - Ciliophora (ciliates) (Chapter 20)
      - Heterokontophyta (Chapter 21)
  - Unikonta (1-flagellum)
    - Amoebozoa (Chapter 22)
    - Opisthokonta
      - Choanozoa (Chapter 23)

J.J. Berman: Taxonomic Guide to Infectious Diseases. DOI: http://dx.doi.org/10.1016/B978-0-12-415895-5.00016-7

Animalia (Chapters 25–32)
Fungi (Chapters 33–37)

Class Metamonda is a class of anaerobic eukaryotes that lack mitochondria. As discussed in Chapter 15 (Overview of Class Eukaryota), the first eukaryotes possessed three defining anatomic structures: a nucleus, mitochondria (one or more), and flagellum (one or more). Evolution can be an intolerant process. Structures that serve no important biological purpose cannot justify the resources required to maintain their continued existence. In particular, mitochondria may be a great way for deriving energy from oxygen, but these complex organelles may have little value when conditions are anaerobic.

As we will see again and again throughout this book, phylogenetic traits are seldom lost, without leaving some trace of their heritage. The anaerobic members of Class Metamonada maintain a relict organelle, derived from an ancestral mitochondrium. The relict is usually referred to as a mitosome, though a specific term, hydrogenosome, is used to refer to mitochondrial relicts that use iron-sulfide proteins to yield molecular hydrogen and ATP. Various so-called amitochondriate eukaryotic classes that have lost classic mitochodria have retained mitosomes or hydrogenosomes that form molecular hydrogen: Class Metamonada (Chapter 16) [65], Class Amoebozoa (Chapter 22) [66], and Class Microsporidia (Chapter 37) [67].

Metamonada
  Trichozoa
    Parabasalidea
      Trichomonadida
        Trichomonadidae
          *Trichomonas
        Monocercomonadidae
          *Dientamoeba

Members of Class Trichomonadida live as commensals with a single morphologic form: trophozoites. Cysts are not formed. The absence of a cyst form is significant because cyst forms resist adverse environmental conditions, permitting the organism to live outside the host for varying lengths of time, and to infect organisms without direct contact with the infected host. Without a cyst form, members of Class Trhichomonadida depend on direct transmission between host organisms.

The two human pathogens in Class Trichomonadida are: *Trichomonas vaginalis* and *Dientamoeba fragilis*.

Trichomonas vaginalis is the most common sexually transmitted disease, with about 8 million new cases occurring annually, in North American [32,70]. Its reservoir is humans. Its role as a causative agent of urethritis, vaginitis, and cervicitis, in women, is well known. Fewer people seem to be aware that both men and women are commonly infected. Though many individuals infected with *Trichomonas vaginalis* are asymptomatic, the organism can cause urethritis in a significant percentage of infected men [71]. Like all members of Class Trichomonadida, it exists only in the trophozoite form, and transmission is typically caused by the exchange of infected secretions during sexual intercourse. The trophozoites can be easily visualized in cervical Pap smears, by light microscopic examination.

Dientamoeba fragilis causes diarrhea and other gastrointestinal symptoms. It causes disease worldwide. Its reservoir seems to be restricted to humans and other primates. Like the other members of Class Trichomonadida, it is found only in the trophozoite form. The organism lives in the large intestine. Though it can be found in stools, it degenerates quickly (hence the "fragilis" in *Dientamoeba fragilis*), and the organism can be difficult to demonstrate. A 2005 study employed a sensitive and specific PCR (polymerase chain reaction) test on stool specimens from 6750 patients with gastrointestinal symptoms and on a control set of 900 asymptomatic individuals. The study found that about 1% of the symptomatic patients tested positive for *Dientamoeba fragilis*. None of the asymptomatic patients demonstrated infection [72]. This would indicate that the organism, when present, causes symptoms; and that Dientamoeba is a significant cause (on the order of 1%) of gastrointestinal complaints. Because the trophozoite is passed in stools, and because the organism has no cyst stage, the presumed method of transmission is through fecal–oral exchange.

Metamonada
  Fornicata
    Diplomonadida
      Hexamitidae
        *Giardia

Members of Class Diplomonadida have a so-called mirror morphology, with two sets of nuclei, flagella, and cytoplasm, symmetrically arranged about a central axis [73]. There is only one human pathogen in Class Diplomonada: *Giardia lamblia* (alternately known as *Giardia intestinalis* or *Giardia duodenalis*, and formerly known as *Lamblia intestinalis*). *Giardia lamblia* can be found worldwide and infects a wide variety of animals, including cats, dogs, and birds. Its infection in beavers has inspired the rhyming couplet, beaver fever. Unlike the pathogens in Class Trichomonadida, Giardia grows in two forms: trophozoite and cyst. The organism lives in the small intestine, in contrast with *Entamoeba histolytica* (Class Amoebozoa, Chapter 22), which lives in the large intestine. *Entamoeba histolytica* and *Giardia lamblia* both have cyst forms, but the two organisms can be distinguished by the morphology of their trophozoites: Giardia trophozoites have flagella; Entamoeba trophozoites do not. The cyst form supports the long-term survival of the organism in contaminated water supplies. Transmission can occur from the fecal–oral route or from ingesting contaminated water. Giardiasis is characterized by diarrhea and associated gastrointestinal complaints.

Infectious species:

Dientamoeba fragilis (dientamoebiasis)
Trichomonas vaginalis (trichomoniasis)
Giardia lamblia, same as *Giardia intestinalis*, *Giardia duodenalis*, and *Lamblia intestinalis* (giardiasis, beaver fever)

Chapter 17

Euglenozoa

"Each organism's environment, for the most part, consists of other organisms."
Kevin Kelly

Eukaryota
  Bikonta (2-flagella)
    Excavata
      Metamonada (Chapter 16)
      Discoba
        Discicristata
          Euglenozoa (Chapter 17)
            Kinetoplastida
              Trypanosomatida
                *Leishmania
                *Trypanosoma
          Percolozoa (Chapter 18)
    Archaeplastida (Chapter 24)
    Chromalveolata (Chapters 19–21)
      Alveolata
        Apicomplexa (Chapter 19)
        Ciliophora (ciliates) (Chapter 20)
      Heterokontophyta (Chapter 21)
  Unikonta (1-flagellum)
    Amoebozoa (Chapter 22)
    Opisthokonta
      Choanozoa (Chapter 23)
      Animalia (Chapters 25–32)
      Fungi (Chapters 33–37)

The Euglenozoa are single-cell organisms that are closely related to the Percolozoans (Chapter 18). Class Euglenozoa plus Class Perocolozoa constitute the only subclasses of Class Discicristata that contain human pathogens. As such, euglenozoans and procolozoans have disc-shaped mitochondria,

J.J. Berman: Taxonomic Guide to Infectious Diseases. DOI: http://dx.doi.org/10.1016/B978-0-12-415895-5.00017-9

characteristic of their superclass. All members of Class Euglenozoa that are pathogenic in humans fall under one class: Class Trypanistomatida. Class Trypanistomatida contains two infectious genera: Leishmania and Trypanosoma, and these two genera account for three of the most debilitating, widespread, and prevalent diseases of humans: leishmaniasis, Chagas disease, and sleeping sickness.

Euglenozoa
  Kinetoplastida
    Trypanosomatida
      *Leishmania
      *Trypanosoma

Class Trypanistomatida is a subclass of Class Kinetoplastida. As members of a subclass of Class Kinetoplastida, members of Class Trypanistomatida contain a unique and perplexing structure known as a kinetoplast. A kinetoplast is a clump of DNA composed of multiple copies of the mitochondrial genome, tucked inside a mitochondrion. All the members of Class Trypanistomatida are parasitic (i.e., they spend much of their lives inside a host organism). The members of class Trypanistomatida that are pathogenic in humans all have a primary host (humans) and an insect serving as an intermediate host.

Members of Class Trypanistomatida have features that are either unique to these organisms or that are seldom found in other organisms:

1. Cell division is neither mitotic nor meiotic in members of Trypanistomatida. The organelle in which the division process is focused is the kinetoplast (a condensed mitochondrial genome). Each cell has one mitochondrion containing one kinetoplast, and the time at which division of the organism occurs is regulated by the cell cycle of the kinetoplast. The kinetoplast divides, while the flagellum, anchored to the kinetoplast at the basal body, replicates, permitting the cell to bifurcate into separate flagellate cells. The nucleus replicates its DNA but never condenses into chromosomes; no mitotic spindle is formed. Nuclear DNA migrates to the new cell when the replicated kinetoplast and flagellum are formed. This features is completely unique to Class Trypanistomatida.
2. Replication occurs in primary and intermediate hosts. In most cases, wherein a parasite has multiple hosts, reproduction occurs in the primary host, and maturation occurs in the secondary host. In the case of members of Class Trypanistomatida, parasites replicate in the primary (animal) and the secondary (insect) hosts.
3. Intracellular forms have flagella. Most flagellate parasites live extracellular lives. Presumably, the flagella makes it hard to live inside a cell. It would be like trying to open your umbrella in a phone booth. Most members of Class Trypanistomatida have a so-called amastigote phase characterized by a very small cell with a tiny, virtually invisible flagellum. Amastigotes are

the intracellular form of the organism. The only disease organism in Class Trypanistomatida that does not have an amastigote phase is *Trypanosoma brucei*, the cause of African trypanosomiasis. This organism has no intracellular phase and is found as a swimming (flagellate) form in blood and body fluids.

4. Organisms have novel methods to control or avoid the host immune response. For example, Leishmania organisms that pass from insect vector (female sandfly; see Glossary item, Sandfly) to human host are quickly phagocytosed by neutrophils. The neutrophils serve as Trojan horses for the leishmania. Normally, after a neutrophil engulfs foreign organisms, the neutrophil soon dies, along with the contained organisms. Leishmania actively stabilizes neutrophils for several days, until macrophages arrive to eat the neutrophil and the leishmania within. The leishmanial organisms live as intracellular organisms within macrophages.

The two infectious genera in Class Trypanistomida are Leishmania and Trypanosoma. Leishmania species cause leishmaniasis, a disease that infects about 12 million people worldwide. Each year, about 60 000 people die from the visceral form of the disease. It is a tropical disease that occurs most often in India, Africa, and Brazil (in the order of decreasing incidence).

Many different Leishmania species infect a wide range of animals, with about 21 different species infecting humans. In most cases, the species that are infective in humans have a non-human animal reservoir.

The target organs, and the subsequent clinical syndrome, vary with the species of Leishmania. All infective species are transmitted by the bite of a female sandfly. There are many species of sandfly, but all of the species that transmit leishmaniasis seem to belong to the Phlebotominae family.

The clinical forms of disease, which are dependent on the infective species of Leishmania, are: cutaneous (localized skin lesion), diffuse cutaneous, mucocutaneous, and visceral (also known as kala-azar). The visceral form of the disease is the most severe form and often results in death, if not treated.

Trypanosoma brucei is the cause of African trypanosomiasis (sleeping sickness). Two subspecies cause infections in humans: *Trypanosoma brucei gambiense* and *Trypanosoma brucei rhodesiense*. The disease occurs almost exclusively in sub-Saharan Africa. The reported numbers of cases are considered to be unreliable, but it has been estimated that infection with *Trypanosoma brucei* accounts for about 50 000 deaths each year. The intermediate host and vector for the disease is the tsetse fly (Genus Glossina), and the reservoir for the tsetse fly is infected humans or animals. As *Trypanosoma brucei* does not have an amastigote stage (capable of intracellular growth), it is found exclusively in extracellular fluids; principally, lymph, blood, and cerebrospinal fluid. Symptoms begin with swollen lymph nodes in the back of the neck, headache, and other clinical features of generalized infection. When the organism crosses the blood–brain barrier, the neurologic phase begins. Symptoms may include irritability, fatigue, insomnia, and irregular somnolence. The organism produces a chemical, tryptophol, known to induce sleep. Untreated cases are often fatal.

Trypanosoma cruzi is the cause of Chagas disease, also known as American trypanosomiasis. Chagas disease occurs almost exclusively in Central and South America. It affects about 8 million people [74]. Its intermediate host and vector is any of several species of the Triatominae subclass of Class Reduviiae insects. These blood-sucking triatomes are sometimes called kissing bugs, as they often bite the face, near the mouth. The triatome bug has a choice of about 100 different animals that serve as reservoirs for *Trypanosoma cruzi*. Infections have been known to occur following transfusion with contaminated blood.

Trypansoma cruzi produces an intracellular amastigote stage, and it is the amastigote that preferentially infects neurons of the peripheral nervous system. Infection of neurons in the wall of the heart can lead to conduction abnormalities, myocarditis, cardiomyopathy, and ventricular aneurysms. Infection of the neurons in the colon can lead to aganglionic megacolon.

Infectious species:

Leishmania aethiopica species complex: *Leishmania aethiopica* (cutaneous leishmaniasis)
Leishmania amazonensis (cutaneous leishmaniasis)
Leishmania brasiliensis (leishmaniasis)
Leishmania chagasi (visceral leishmaniasis)
Leishmania donovani (visceral leishmaniasis)
Leishmania infantum (visceral leishmaniasis)
Leishmania major (cutaneous leishmaniasis)
Leishmania mexicana (cutaneous leishmaniasis)
Leishmania peruviana (leishmaniasis)
Leishmania tropica (cutaneous leishmaniasis)
Leishmania venezualensis (cutaneous leishmaniasis)
Leishmania Viannia *brasiliensis* (leishmaniasis)
Leishmania Viannia *guyayensis* (leishmaniasis)
Leishmania Viannia *panamensis* (leishmaniasis)
Trypanosoma cruzi (American trypanosomiasis, Chagas disease)
Trypanosoma brucei gambiense (African trypanosomiasis, sleeping sickness)
Trypanosoma brucei rhodesiense (African trypanosomiasis, sleeping sickness)

Chapter 18

Percolozoa

"The cell is basically an historical document."

Carl R. Woese [21]

- Eukaryota
  - Bikonta (2-flagella)
    - Excavata
      - Metamonada (Chapter 16)
      - Discoba
        - Euglenozoa (Chapter 17)
        - Percolozoa (Chapter 18)
          - Heterolobosea
            - Schizopyrenida
              - Vahlkampfiidae
                - *Naegleria
    - Archaeplastida (Chapter 24)
    - Chromalveolata (Chapters 19–21)
      - Alveolata
        - Apicomplexa (Chapter 19)
        - Ciliophora (ciliates) (Chapter 20)
      - Heterokontophyta (Chapter 21)
  - Unikonta (1-flagellum)
    - Amoebozoa (Chapter 22)
    - Opisthokonta
      - Choanozoa (Chapter 23)
      - Animalia (Chapters 25–32)
      - Fungi (Chapters 33–37)

Below Class Bikonta (two flagella) is Class Excavata (from Latin excavare, to make hollow, or, in this case, grooved). All subclasses of Class Excavata descend from an organism with a ventral feeding groove. Under Class Excavata is Class Discoba.

J.J. Berman: Taxonomic Guide to Infectious Diseases. DOI: http://dx.doi.org/10.1016/B978-0-12-415895-5.00018-0

Class Discoba is a newly invented class, and its taxonomic origin is instructive. Recently acquired molecular data suggest that three subclasses of Class Excavata share a common direct ancestor: Class Percoloza (Chapter 18), Class Euglenozoa (Chapter 17), and Class Jakobid (which happens to contain no infectious organisms). The common ancestor of these three classes is apparently not shared with another subclass of Excavata: Class Metamonada (Chapter 16). To preserve monophyly within Class Excavata, a newly named class needed to be inserted under Class Excavata. This class would contain Class Percolozoa, Class Euglenozoa, and Class Jakobid and would exclude Class Metamonada. The newly named class is Discoba [75].

Under Class Discoba is Class Percolozoa, single-celled organisms containing mitochondria with discoid cristae, and the ability to shift between three morphologic forms: amoeboid, flagellate, and cyst. The amoeboid form consists of a non-flagellated feeding cell. Like all amoeboid forms, it moves slowly, by extending a section of its cytoplasm (the pseudopod). Under adverse conditions, the amoeboid form can develop flagella, which presumably enhance its ability to move to a more hospitable location. As you would expect from organisms in a subclass of Bikonta, two flagella are observed in the flagellate form. Under conditions that are severely unsuitable for growth, the organism converts to a cyst form.

Percolozoa
  Heterolobosea
    Schizopyrenida
      Vahlkampfiidae
        *Naegleria

There is only one known pathogenic genus in Class Percolozoa: Naegleria. *Naegleria fowleri* is the only species of Naegleria that is known to be infectious in humans. In older microbiology textbooks, Naegleria is grouped with the Acanthamoeba, under Class Amoebozoa (Chapter 22). Naegleria rightly belongs in Class Percolozoa. The naeglerian life cycle includes three stages: cyst, trophozoite (amoeboid), and flagellate forms; whereas pathogenic members of Class Amoebozoa have only two stages: cyst and trophozoite. This distinction has taxonomic and diagnostic relevance. Though naeglerian meningoencephalitis can be confused histologically with amoebic meningoencephalitis, flagellate forms in cerebrospinal fluid would indicate a naeglerian infection (not an amoebic infection).

Naegleria fowleri is found in fresh water and in poorly chlorinated swimming pools, where it is free-living. Infections result from exposure to free-living organisms in their natural habitat, and not from exposure to infected individuals. The organism travels from the nose to the brain, where it causes a meningoencephalitis. Because these organisms are widely found in water, it is presumed that millions of people are exposed to the organism, but only rare individuals develop meningoencephalitis. It is not known why most people are unaffected by the organism, while others develop a rapidly progressive meningoencephalitis, with a very high mortality rate (about 95%). Naegleria meningoencephalitis accounts for several deaths in the United States of America each year. Several recent cases of Percolozoan meningoencephalitis have occurred in individuals who used unsterilized water in Neti pots to irrigate their nasal sinuses.

Infectious species:

Naegleria fowleri (meningoencephalitis)

Chapter 19

Apicomplexa

"The world appears to me to be put together in such a painful way that I prefer to believe that it was not created . . . intentionally."

Stanislaw Lem

Eukaryota
  Bikonta (2-flagella)
    Excavata
      Metamonada (Chapter 16)
      Discoba
        Euglenozoa (Chapter 17)
        Percolozoa (Chapter 18)
    Archaeplastida (Chapter 24)
    Chromalveolata (Chapters 19–21)
      Alveolata
        Apicomplexa (Chapter 19)
          Aconoidasida
            Haemosporida
              Plasmodiidae
                *Plasmodium
            Piroplasmida
              Babesiidae
                *Babesia
          Conoidasida
            Coccidia
              Eucoccidiorida
                Eimeriorina
                  Cryptosporidiidae
                    *Cryptosporidium
                  Eimeriidae
                    *Cyclospora
                    *Isospora (same as Cystoisospora)
                Sarcocystidae

J.J. Berman: Taxonomic Guide to Infectious Diseases. DOI: http://dx.doi.org/10.1016/B978-0-12-415895-5.00019-2

*Sarcocystis
*Toxoplasma
Ciliophora (ciliates) (Chapter 20)
Heterokontophyta (Chapter 21)
Unikonta (1-flagellum)
Amoebozoa (Chapter 22)
Opisthokonta
Choanozoa (Chapter 23)
Animalia (Chapters 25–32)
Fungi (Chapters 33–37)

Apicomplexa and Ciliophora (Chapter 20) are the two classes within Class Alveolata that contain organisms that produce human disease. Members of Class Alveolata are single-celled organisms with distinctive sacs underlying the plasma membrane (alveoli). They all have characteristic mitochondria, containing tubular cristae.

Members of Class Apicomplexa are all spore-forming parasites that live in animals. They all have a characteristic structure on the apex of their cells that helps them invade into animal cells; hence, the name Apicomplexa, referring to the apical complex. The lives of the apicomplexans are complex and individualized, but all members share several basic properties: the ability of apicomplexan cells to leave host cells, to travel to other host cells, to invade host cells, to feed and grow within host cells, and to divide into many small cells through a process called schizogony.

Schizogony is a biological feature that characterizes the apicomplexans, accounting for most of the histopathologic findings in apicomplexan infections, and inspiring a set of arcane terms that apply specifically to apicomplexans. When you understand schizogony, you understand apicomplexan pathology.

When human cells divide, they undergo a long process wherein the cell produces all of the material required for two full-sized cells, before it splits: two sets DNA, double the volume of cytoplasm, double the nuclear proteins, double the number of mitochondria, and a plasma membrane sufficient to cover two full-sized cells. In humans, cell division requires much energy and time. The apicomplexans have discovered a short-cut that greatly reduces the time and energy required for cell division. The shortcut is called schizogony, and consists of rapid nuclear replication without synchronous cytoplasmic synthesis. Basically, the apicomplexan cell can divide into a large number of much smaller cells, and these smaller cells serve a specific purpose in the apicomplexan cell cycle. The two apicomplexan cell types that are capable of schizogony are the trophozoite and the oocyst.

The full-sized, vegetative apicomplexan cell is the trophozoite. It lives inside a host cell (such an erythrocyte or a hepatocyte, or an intestinal epithelial cell), feeding off the host. After a time, the trophozoite undergoes schizogony, producing a collection of small, infective cells called merozoites, that travel (extracellularly) to other host cells, where they invade, feed, multiply, and grow into trophozoites within specific target cells. The trophozoite → merozoite cycle continues for a time, producing large numbers of infectious cells within the host organism. Eventually, the merozoites enter a new stage of life cycle stage, as male or female gametes. Depending on the species of Class Apicomplexa, fertilization may take place in the original host or in another organism (the female Anopheles mosquito in the case of malaria). The fertilized egg develops into an oocyst (that can survive outside the host). Under favorable conditions, the oocyst will undergo schizogony to produce small sporozoites that infect cells within a new host. Like the trophozoite, the sporozoite produces merozoites. Members of Class Apicomplexa have life cycles that vary somewhat from this description, and interested readers are encouraged to study the life cycles of the individual species. Nonetheless, knowledge of the most generic apicomplexan life cycle will help students understand the basic biology of the many important diseases produced by the thousands of different apicomplexan species that infect animals.

Seven genera of Class Apicoomplexa cause diseases in humans: Plasmodium, Babesia, Cryptosporidium, Cyclospora, Cystoisospora, Sarcocystis, and Toxoplasma. Plasmodium and Babesia are subclasses of Class Aconoidasida (apicomplexans that lack a cone at the tip). The other apicomplexans are subclasses of Class Conoidasida (apicomplexans that have a cone at the tip).

Apicomplexa
- Aconoidasida
  - Haemosporida
    - Plasmodiidae
      - *Plasmodium
  - Piroplasmida
    - Babesiidae
      - *Babesia

Genus Plasmodium is responsible for human and animal malaria. About 300–500 million people are infected with malaria, worldwide. About 2 million people die each year from malaria [4,76]. Malaria accounts for more human deaths than any other vector-borne illness. There are several hundred species of Plasmodium that infect animals, but only a half dozen species are known to infect humans [76]. Newly discovered species, causing human disease, may arise from animal reservoirs. For example, *Plasmodium knowlesi* causes malaria in macaque monkeys. It has emerged as a human pathogen in Southeast Asia, where it currently accounts for about two-thirds of malarial cases.

In humans, malaria is contracted from sporozoites injected by female Anopheles mosquitoes. The sporozoites eventually develop into merozoites.

In some species (including *Plasmodium ovale*, *Plasmodium vivax*, but not in *Plasmodium malariae*) a dormant variant of merozoite is produced that can produce a malarial relapse after decades of remission [77].

The Genus Plasmodium has some very distinctive features, not the least of which is its uniquely low G+C ratio. *Plasmodium falciparum* has a G+C ratio of about 20%, much lower than the ratios seen in the so-called low G+C bacteria (Class Bacilli and Class Clostridia, Chapter 12), that hover just under 50%.

Human babesiosis is uncommon, but this organism infects many mammals other than humans, and is the second most common parasitic blood disease of non-human mammals (after trypanosomal infections).

Two species infect humans: *Babesia divergens*, the most common cause of babesiosis in Europe, and *Babesia microti*, the most common type of babesiosis in North America [78]. The tick, Genus Ixodidae, serves an almost identical role for Genus Babesia as the Anopheles mosquito serves for Genus Plasmodium: sporozoites develop in the salivary gland of the tick and are introduced into the human host, via a bite. In babesiosis, unlike malaria, infection occurs exclusively in red cells and is not found in liver cells. Aside from humans, animal reservoirs of *Babesia microti* and *Babesia divergens* include mice and deer. Like malaria, babesiosis can be transmitted via transfusions with contaminated blood products (see Glossary item, Blood contamination).

Apicomplexa
    Conoidasida
        Coccidia
            Eucoccidiorida
                Eimeriorina
                    Cryptosporidiidae
                        *Cryptosporidium
                    Eimeriidae
                        *Cyclospora
                        *Isospora (same as Cystoisospora)
                Sarcocystidae
                    *Sarcocystis
                    *Toxoplasma

Six species of Class Coccidia produce disease in humans: *Cryptosporidium parvum*, *Cyclospora cayetanensis*, *Cystoisospora belli* (*Isospora belli*), *Sarcocystis suihominis*, *Sarcocystis bovihominis*, and *Toxoplasma gondii*.

As a subclass of Class Apicomplexa, the members of Class Coccidia are obligate intracellular parasites. Nonetheless, Class Coccidia confers two features that make it possible for coccidan species to survive outside the host, for a period of time. First, there is the durable coccidian oocyst. Most coccidians attach to enterocytes (particularly, the epithelial cells that line the small intestine). From this location, oocysts are carried within feces and eventually contaminate the water where feces are deposited. One method of coccidian infection is through oocysts that contaminate food and water. A second strategy for long-term survival is achieved by some coccidians that encyst within the tissues of the host organism. This coccidian cyst is formed from the husk of a host cell that has been filled by multiplying small forms of the organism. These so-called cysts, which are really modified host cells filled with infective organisms, are stable and durable. When the host tissues are eaten by another animal, the cysts release an infective form of the organism, and the coccidian life cycle repeats, beginning in the intestines.

The disease produced by any of the organisms belonging to Class Coccidia is known collectively as coccidiosis. This term is applied most often to coccidian infections in animals. Coccidiosis must not be confused with the similar-sounding coccidioidomycosis (Ascomycota, Chapter 36).

Cryptosporidiosis is caused by *Cryptosporidium parvum*. It is typically transmitted by drinking water that has been contaminated with relatively stable oocysts carried by the feces of an infected animal. Infection is confined to the mucosa of the small intestine, typically producing a watery diarrhea that is self-limited in immune-competent individuals. Cryptosporidiosis infection can produce a severe gastroenteritis in immune-deficient individuals. In some cases of Cryptosporidiosis occurring in immune-deficient persons, the organism is never completely cleared from the small intestine. Cryptosporidiosis is a common disease that occurs worldwide.

Like all members of Class Apicomplexa, Genus Cryptosporidium is an obligate intracellular organism. Casual histologic examination of infected mucosa shows that the organisms perch atop the apical surface of the epithelial cells lining the small intestinal villi (i.e., extracellular growth). Careful examination demonstrates that the cryptosporidia attached to the mucosal surface are actually wrapped by host cell membranes (i.e., intracellular growth). Students should avoid confusing cryptosporidiosis with the similar-sounding cryptococcosis (Class Basidiomycota, Chapter 35).

One species of Genus Cyclospora produces disease in humans: Cyclospora cayetanensis. As in cryptosporidiosis, stable oocysts carried in water contaminated by infected animals, pass into the gastrointestinal tract and produce sporozoites that grow in the small intestine. The infection produces diarrhea. On histologic stains, *Cyclospora cayetanensis* can be mistaken for *Cryptosporidium parvum*, but *Cyclospora cayetanensis* is the larger of the two organisms.

Cystoisospora (synonymous with *Isospora*) *belli* has a life cycle, mode of transmission, and clinical presentation (enteritis), similar to that of *Cryptosporidium parvum* and *Cyclospora cayetanensisis*. Isosporiasis is an uncommon disease, but when it occurs, it tends to arise in immune-deficient individuals. Diagnosis can usually be made after careful stool examination. The oocysts are ellipsoidal and large.

Cryptosporidium, Cyclospora, and Cystoisospora, the three enteric coccidians, are genera under Class Eimeriorina. Two of the coccidians are genera under Class Sarcocystidae: Sarcocystis and Toxoplasma. These genera produce a specialized structure, containing dormant, infectious organisms, filling the husk of a host cell. These structures have specialized names, but most healthcare workers use the short but inaccurate term, "cysts."

Sarcocystis suihominis and *Sarcocystis bovihominis* can cause disease in humans. As with other coccidians, oocysts are released into the gut lumen and pass into the environment via feces. The intermediate host ingests contaminated water, and infection leads to the development of cysts in host cells, usually muscle; hence the name sarcocyst (from the Greek, "sarx," flesh). When the intermediate host is eaten by a primary host, the organisms within the sarcocyst reproduce, and their oocytes are eventually released into the gut, where they mix with fecal material and leave the host. The sarcocyst's primary host has a purely enteric infection. The intermediate host, in which ingested organisms invade muscle throughout the body, and develop sarcocysts within muscle cells, has an enteric and a systemic disease. As their names would suggest, infections occur in swine (hence, suihominis), or cows (hence, bovihominis). On rare occasions, humans become infected and may serve as either primary or secondary hosts. Though sarcosporidiosis is considered a rare disease in humans, the animal reservoir for human infections is large. Various species of Genus Sarcocystis commonly infect a wide range of animals worldwide. In Southeast Asia, post-mortem biopsies of human tongues revealed sarcocysts in 21% of the sampled population [79]. This would indicate that many human infections are asymptomatic.

The sister genus to Sarcocystis is Toxoplasma. About one third of the human population has been infected (i.e., about 2.3 billion people) by the only species that produces human toxoplasmosis: *Toxoplasma gondii*. This number is somewhat higher than the number of tuberculosis infections worldwide (about 2 billion). In the United States of America, the prevalence rate of Toxoplasma infection is about 11%.

Toxoplasma has a life cycle very similar to Sarcocystis. The most common primary host of Toxoplasma is the cat. Humans are the intermediate host. For most infected individuals, infection does not cause overt disease. Serious disease occurs most often in immune-compromised hosts, particularly individuals with AIDS or individuals who are pregnant. Transplacental infections may occur. Disease occurs in the locations where the cysts develop, often in the brain (causing encephalitis), or eyes (causing chorioretinitis), or through which the organisms migrate (e.g., lymph nodes). Latent cysts can produced a reactivation of toxoplasmosis, and this occurs most often in individuals who were infected early in life and later became immune-compromised due to disease or immunosuppressive therapy.

Infectious species:

Plasmodium falciparum (tertian malaria)
Plasmodium vivax (tertian malaria)
Plasmodium ovale (tertian malaria)
Plasmodium malariae (quartan malaria)
Plasmodium knowlesi (zoonotic malaria)
Babesia divergens (babesiosis, piroplasmosis)
Babesia microti (babesiosis, piroplasmosis) (the most common type of babesiosis in North America [78])
Babesia duncani (babesiosis, piroplasmosis)
Cryptosporidium parvum (cryptosporidiosis)
Cyclospora cayetanensis (cyclosporiasis, gastroenteritis, some cases of traveler's diarrhea)
Cystoisospora belli, formerly and still used, *Isospora belli* (isosporiasis)
Sarcocystis suihominis, formerly *Isospora hominis* (sarcocystosis, sarcosporidiosis)
Sarcocystis bovihominis, also known as *Sarcocystis hominis*, formerly *Isospora hominis* (sarcocystosis, sarcosporidiosis)
Toxoplasma gondii (toxoplasmosis)

Chapter 20

Ciliophora (Ciliates)

"There should be some things we don't name, just so we can sit around all day and wonder what they are."

George Carlin

- Eukaryota
  - Bikonta (2-flagella)
    - Excavata
      - Metamonada (Chapter 16)
      - Discoba
        - Euglenozoa (Chapter 17)
        - Percolozoa (Chapter 18)
    - Archaeplastida (Chapter 24)
    - Chromalveolata (Chapters 19–21)
      - Alveolata
        - Apicomplexa (Chapter 19)
        - Ciliophora (ciliates) (Chapter 20)
          - Litostomatea
            - Vestibuliferida
              - Balantiididae
                - *Balantidium
      - Heterokontophyta (Chapter 21)
  - Unikonta (1-flagellum)
    - Amoebozoa (Chapter 22)
    - Opisthokonta
      - Choanozoa (Chapter 23)
      - Animalia (Chapters 25–32)
      - Fungi (Chapters 33–37)

Ciliophora and Apicomplexa (Chapter 19) are the two subclasses of Class Alveolata that contain human pathogens. Members of Class Alveolata are single-celled organisms with distinctive sacs underlying the plasma membrane (alveoli). The mitochondria of alveolates have tubular cristae (internal partitions).

J.J. Berman: Taxonomic Guide to Infectious Diseases. DOI: http://dx.doi.org/10.1016/B978-0-12-415895-5.00020-9

Members of Class Ciliophora have a peculiar system of two nuclei: a small nucleus that contains the full genome, and a larger nucleus that plays the active role in cellular regulation. The larger nucleus is regenerated from the smaller nucleus through a process of gene amplification and gene editing. This produces a functional, but non-reproductive nucleus, with an enrichment of genes that play a role in cell regulation, and a reduction of genes that are involved in nuclear regeneration. It would seem that the members of Class Ciliophora have attained a mechanism whereby the nucleus divides its labors by creating a back-up nucleus that preserves a full set of genes, and a working nucleus, with a gene ensemble selected for cellular functionality.

Ciliophora
  Litostomatea
    Vestibuliferida
      Balantiididae
        *Balantidium

There is only one species of Class Ciliophora that is infectious in humans: *Balantidium coli*. Balantidiasis is a zoonosis, found in a variety of animals, but the most important reservoir is the pig. In pigs, the organisms produce no apparent clinical disease. The organisms live in the pig's colon, and are passed into the feces. Human disease, which is rare, occurs when humans ingest food that has been contaminated by pig feces, or, rarely, from the feces of infected humans.

There are two forms of the organism: trophozoite and cyst. The trophozoites are the active, growing form of the organisms. Trophozoites transform into cysts under poor growth conditions. Encystation typically occurs in the distal colon, where the feces are relatively dry, and unfavorable for growth. The cysts are inactive, but can survive adverse conditions. Cysts are passed in the feces and are the passively infective form of the organism. Once in the intestine, cysts transform into trophozoites.

The trophozoites are capable of invading into the wall of the colon and occasionally produce ulcers. The ulcers of balantidiasis have a peculiar inverted-flask shape (i.e., a small opening on the lumen surface overlying edges that extends laterally and ends sharply as a flat base, in the colon wall). Most cases of human balantidiasis are mild (some diarrhea or constipation) or asymptomatic. Cases of balantidiasis occur worldwide, but they are rare in the USA. Balantidiasis is a common infection in the Philippines.

Infectious species:

Balantidium coli (balantidiasis)

Chapter 21

Heterokontophyta

"Biology is the science. Evolution is the concept that makes biology unique."
Jared Diamond

- Eukaryota
  - Bikonta (2-flagella)
    - Excavata
      - Metamonada (Chapter 16)
      - Discoba
        - Euglenozoa (Chapter 17)
        - Percolozoa (Chapter 18)
    - Archaeplastida (Chapter 24)
    - Chromalveolata (Chapters 19–21)
      - Alveolata
        - Apicomplexa (Chapter 19)
        - Ciliophora (ciliates) (Chapter 20)
      - Heterokontophyta (Chapter 21)
        - Blastocystae
          - Blastocystida
            - Blastocystidae
              - *Blastocystis
  - Unikonta (1-flagellum)
    - Amoebozoa (Chapter 22)
    - Opisthokonta
      - Choanozoa (Chapter 23)
      - Animalia (Chapters 25–32)
      - Fungi (Chapters 33–37)

Heterokontophyta (also known as stramenopiles) are bikonts (two flagella). The name "heterokont" derives from a characteristic morphologic feature; the two flagella are non-identical. The class contains two strikingly different types of organisms; a colored group containing various types of algae and

J.J. Berman: Taxonomic Guide to Infectious Diseases. DOI: http://dx.doi.org/10.1016/B978-0-12-415895-5.00021-0

diatoms, and a colorless group containing organisms that have morphologic features similar to fungi.

Class Heterokontophyta has undergone extensive taxonomic revision in the recent past. The golden and brown algae (currently in Class Heterokontophyta) were previously classified under Class Plantae. Adding to the confusion, green and red algae are currently placed in Class Archaeplastida (Chapter 24). Various species of the colorless group (currently in Class Heterokontophyta) had been incorrectly placed in Class Fungi. Oomycota, a "colorless" class of heterokonts, contains the organisms that produce late blight of potato (*Phytophthora infestans*), and sudden oak death (*Phytophthora ramorum*). Oomycota, despite its suffix (mycota, an alternative name for fungus), is not a member of Class Fungi.

Given the taxonomic turmoil within Class Heterokontophya, students should be grateful that there is only one heterokont genus that is infectious in humans: Blastocystis.

Heterokontophyta
  Blastocystae
    Blastocystida
      Blastocystidae
        *Blastocystis

Blastocystis infects a variety of animals. The general rule for naming species of Blastocystis is to apply the genus to the species in which the organisms are recovered. Hence, multiple species of Blastocystis recovered from human feces were named *Blastocystis hominis*. Species of the organism recovered from rat feces were all named *Blastocystis ratti*. Obviously, this taxonomic disaster cannot persist. Efforts to identify individual species by species-specific genotyping (rather than bundling different species under the host species name) have begun [80].

Infection seems to be through acquisition of the cyst form of the organism, through a fecal–oral route. Human–human, animal–human, and human–animal transmission all may occur, but the relative frequencies of these different transmissions are unknown. Incidence is highest where humans are exposed to animal feces, implying that animal–human transmission is common. In the USA the prevalence rate of infection of *Blastocystis hominis* is 23%, with the highest rates found in the western states [81].

Four different morphologic forms of Blastocystis have been observed: cyst, vacuolar, amoeboid, and granular. After the cyst is ingested, the other forms emerge in the intestine. Replication of the organism occurs in the vacuolar form. The length of infection varies from weeks to years. Many Blastocystis infections do not manifest clinically, and an asymptomatic carrier state is common. Disease, when it occurs, can closely mimic irritable bowel syndrome. When species of *Blastocystis hominis* are identified by type-specific laboratory testing, we will be in a better position to determine whether specific variants of the organism are correlated with clinical expression (e.g., asymptomatic disease, long-term carrier state, irritable bowel-like symptoms, inflammatory bowel disease symptoms).

As a final point, it is important to avoid confusing Blastocystis with two similar terms appearing elsewhere in this book: blastomycosis and blastocyst. Blastomycoses is a fungal disease (Ascomycota, Chapter 36), and blastocyst is the hollowed embryonic body characteristic of animals (Overview of Kingdom Animalia, Chapter 25). All three terms come from the root word blastos (Greek for bud or embryo). Cystos (as in Blastocystis and blastocyst) is the Greek root meaning sac.

Infectious species:

Blastocystis hominis (enteritis)

Chapter 22

Amoebozoa

"Because all of biology is connected, one can often make a breakthrough with an organism that exaggerates a particular phenomenon, and later explore the generality."

Thomas R. Cech

- Eukaryota
  - Bikonta (2-flagella)
    - Excavata
      - Metamonada (Chapter 16)
      - Discoba
        - Euglenozoa (Chapter 17)
        - Percolozoa (Chapter 18)
    - Archaeplastida (Chapter 24)
    - Chromalveolata (Chapters 19–21)
      - Alveolata
        - Apicomplexa (Chapter 19)
        - Ciliophora (ciliates) (Chapter 20)
      - Heterokontophyta (Chapter 21)
  - Unikonta (1-flagellum)
    - Amoebozoa (Chapter 22)
      - Lobosea
        - Discosea
          - Thecamoebidae
            - *Sappinia
        - Centramoebida
          - Acanthamoebidae
            - *Acanthamoeba
            - *Balamuthia
      - Conosa
        - Archamoebae
          - Pelobiontida
            - *Entamoeba

J.J. Berman: Taxonomic Guide to Infectious Diseases. DOI: http://dx.doi.org/10.1016/B978-0-12-415895-5.00022-2

Opisthokonta
  Choanozoa (Chapter 23)
  Animalia (Chapters 25–32)
  Fungi (Chapters 33–37)

Class Eukaryota is broadly divided into two large domains, the Bikonta (two flagella) and the Unikonta (one flagellum). The Amoebozoa belong to the Class Unikonta. Most extant Amoebozoa species have no flagella, but a few species of amoebozoans retain their ancestral flagellum. Because the amoebozoans are Unikonts, they are related to Class Opisthokonta (Chapters 23, 25–37), which includes Class Choanozoa, Class Animalia, and Class Fungi.

Amoebozoa move by flowing their cytoplasm from one area of the cell to another. Movement begins when a part of the cell wall, the lobopodium, is protruded. As cytoplasm flows into the lobopodium, the rest of the cell cannot help but follow. The amoebozoan genera vary greatly in size. Species producing disease in humans are about the size of human cells (10–40 microns), while at least one non-pathogenic species, *Amoebozoa proteus*, attains a size of 800 microns (on the verge of visibility with the unaided eye). The members of Class Amoebozoa engulf and eat smaller organisms. Most members of Class Amoebozoa live in the soil or aquatic environments, where they are beneficial bacterial predators (Entamoeba is an exception, see below) [73]. The pathogenic members of Class Amoebozoa occur in tissues as trophozoites (the amoeboid feeding cells) or as cysts (round, infective cells resistant to desiccation).

Determining the major subclasses of Class Amoebozoa has proven to be very difficult [82]. It is likely that the class schema shown in this chapter will be changed in the next few years, when newly acquired genomic information is studied along with previously documented morphologic and physiologic data. In this book, Class Amoebozoa is divided into two major subclasses: Class Lobosea and Class Conosa. All of the pathogenic genera within Class Lobosea are capable of producing meningoencephalitis. Class Conosa includes one pathogenic genus: *Entamoeba histolytica*, which typically produces a dysentery-like condition. Aside from this division that neatly distinguishes the primary infections of the central nervous system and the primary infections of the intestines, the taxonomy of the infectious Amoebozoans is best studied genus-by-genus.

Amoebozoa
  Lobosea
    Discosea
      Thecamoebidae
        *Sappinia
    Centramoebida
      Acanthamoebidae
        *Acanthamoeba
        *Balamuthia

Sappinia is a genus of free-living organisms, within Class Lobosea. There are two species capable of producing meningoencephalitis: *Sappinia diploidea* and its nearly identical species, *Sappinia pedata*. Sappinia species can be found in trophozoite form (i.e., feeding amoeboid forms) and cyst form. The trophozoites are recognized by their two closely apposed nuclei, with a central flattening. Sappinia encephalitis may occur in immune-competent individuals. Only a few cases have been reported [83].

Genus Acanthamoeba and Genus Balamuthia are members of Class Acanthamoebidae. There are a variety of Acanthamoeba species that cause a similar set of human diseases. These include: *Acanthamoeba rhysoides*, *Acanthamoeba polyphaga*, *Acanthamoeba palestinensis*, *Acanthamoeba hatchetti*, *Acanthamoeba culbertsoni*, *Acanthamoeba castellani*, and *Acanthamoeba astronyxis*. These similar species are herein aggregated under the name Acanthamoeba species.

The Acanthamoeba are ubiquitous and are found in the domestic water supply. Acanthamoeba species cause three distinctive clinical diseases: granulomatous amoebic encephalitis, amoebic keratitis, and cutaneous acanthamoebiasis [83]. Though many people are exposed to Acanthamoeba species, few develop disease. Granulomatous encaphalitis and cutaneous acanthamoebiasis occur most often in people with immune deficiencies. Trophozoites and cyst forms can be observed in infected tissues. The trophozoite form may be confused histologically with macrophages. Trophozoites have a single nucleus with a single large nucleolus. The cyst form can be visualized with a variety of histologic stains [83]. Acanthamoebic granulomatous encephalitis is a very rare disease, with a few hundred cases reported, worldwide. Only a few cases of cutaneous acanthamoebiasis have been reported.

Acanthamoebic keratitis can occur in immune-competent individuals. It occurs most often in contact lens users, presumably from washing the lens in water containing the Acanthamoeba species.

The other pathogenic genus within Class Acanthamoebidae is Balamuthia. The only known pathogenic species of this genus is *Balamuthia mandrillaris*, which causes a granulomatous encephalitis, much like that caused by Acanthamoeba species. The organism is found in soil, and possibly water. Like the other pathogenic members of Class Amoebozoa, it occurs in two forms: trophozoite and cyst. In tissue sections, *Balamuthia mandrillaris* is indistinguishable from Acanthamoeba species by standard microscopic examination. Cases of Balamuthia encephalitis are extremely rare, with only a few cases reported in the USA each year, primarily in immune-compromised patients.

Amoebozoa
- Conosa
  - Archamoebae
    - Pelobiontida
      - *Entamoeba

Genus Entamoeba contains the most commonly occurring pathogenic amoebic species, *Entamoeba histolytica* [73]. Infection occurs through fecal–oral transmission of the cyst form. It is estimated that about 50 million people are infected by *Entamoeba histolytica,* with about 70 000 deaths per year, worldwide. All of the other amoebozoan organisms discussed in this chapter, combined, account for just a few dozen deaths. Most *Entamoeba histolytica* infections do not result in clinical disease. The organism can live in the intestine for years as a commensal. Unlike the pathogenic amoebas discussed so far, *Entamoeba histolytica* is not found widely distributed in soil samples. *Entamoeba histolytica* has adapted itself to life inside a narrow range of preferred hosts: primates. As such, its life is restricted to the GI tract of humans and other primates, and to their feces. In the GI tract, the organism can be found in its amoeboid feeding form (i.e. Trophozoite). In stools, the trophozoites encyst; cysts can survive for months.

Entamoeba gingivalis is a species of Entamoeba that is routinely found in a high percentage of specimens obtained from human saliva or from gingival scrapings. Its pathogenicity in oral diseases is controversial. Surprisingly, a recent case report has demonstrated *Entamoeba gingivalis* in a pulmonary abscess [84].

Entamoeba comes with four taxonomic traps that have been used to mortify generations of students.

1. First and foremost, do not confuse *Entamoeba coli* (abbreviation *E. coli*) with *Escherichia coli* (likewise abbreviated as *E. coli*). Both live in the colon, and both can be reported in stool specimens.
2. Of the Entamoeba genera that can be found on examination of human stools, only one is frequently pathogenic, *Entamoeba histolytica. Entamoeba dispar*, *Entamoeba hartmanni*, *Entamoeba coli*, *Entamoeba moshkovskii*, *Endolimax nana*, and *Iodamoeba butschlii*, occasionally found in stool specimens, are generally non-pathogenic.
3. Do not confuse Entamoeba (Class Amoebozoa, Chapter 22) with Dientamoeba (Class Metamonada, Chapter 16).
4. Terminology for amoebic infections is somewhat confusing. It is commonly agreed that the term "amoebiasis," with no qualifiers in the name, refers exclusively to the intestinal infection by *Entamoeba histolytica*, and is also known by the name "amoebic dysentery." Encephalitides caused by members of Class Amoebozoa (Acanthamoeba and Balamuthia) are named granulomatous amoebic encephalitis. Encephalitis caused by *Naegleria fowleri* (not an amoeba) is called primary amoebic meningoencephalitis, an accepted misnomer. Naegleria is a member of class Percolozoa (Chapter 18). A better name for the Naeglerian disease would be primary percolozoan meningoencephalitis. *Acanthamoeba castellanii* causes amoebic keratitis, and is an occasional cause of

granulomatous amoebic encephalitis. *Sappinia diploidea* causes Sappinia amoebic encephalitis.

Infectious species:

Acanthamoeba species (granulomatous amoebic encephalitis, amoebic keratitis, cutaneous acanthamoebiasis)
Balamuthia mandrillaris (granulomatous amoebic encephalitis)
Entamoeba histolytica (amoebic dysentery, amoebic colitis, amoebic liver abscess, amoebic brain abscess, entamoebiasis, amoebiasis)
Entamoeba gingivalis (periodontal disease)
Sappinia diploidea (Sappinia amoebic encephalitis)
Sappinia pedata (Sappinia amoebic encephalitis)

Chapter 23

Choanozoa

"Half of being smart is knowing what you are dumb about."

Solomon Short

Eukaryota
  Bikonta (2-flagella)
    Excavata
      Metamonada (Chapter 16)
      Discoba
        Euglenozoa (Chapter 17)
        Percolozoa (Chapter 18)
    Archaeplastida (Chapter 24)
    Chromalveolata (Chapters 19–21)
      Alveolata
        Apicomplexa (Chapter 19)
        Ciliophora (ciliates) (Chapter 20)
      Heterokontophyta (Chapter 21)
  Unikonta (1-flagellum)
    Amoebozoa (Chapter 22)
    Opisthokonta
      Choanozoa (Chapter 23)
        Ichthyosporea
          Dermocystida
            *Rhinosporidium
      Animalia (Chapters 25–32)
      Fungi (Chapters 33–37)

Class Choanozoa, also known as Class Holozoa, is a subclass of Class Opisthokonta, protists with a single, posterior flagellum. The precise hierarchical position of Class Choanaozoa is subject to change. In the schema below, Class Choanozoa has the same rank as Class Animalia, but it may soon shift leftward, indicating that the Choanozoa are the direct ancestral

J.J. Berman: Taxonomic Guide to Infectious Diseases. DOI: http://dx.doi.org/10.1016/B978-0-12-415895-5.00023-4

class of Class Animalia. Class Choanozoa contains one genus, with one species that is pathogenic in humans: *Rhinosporidium seeberi*.

Choanozoa
  Ichthyosporea (same as Mesomycetozoea)
    Dermocystida
      *Rhinosporidium

Rhinosporidiosis presents clinically as a mass growing on the nasal respiratory lining, often causing nasal obstruction. These growths may occur on other mucosal surfaces, including the ocular conjunctivae, and can occur in animals other than humans. The diagnosis is usually established by histologic examination and by clinical presentation. Tissue sections are characterized by inflammatory tissue embedding large round organisms (the trophocytes).

Members of Class Choanozoa are aquatic, and cases of human rhinosporidiosis can usually be associated with exposure to water in ponds, lakes, or rivers. However, the organism, *Rhinosporidium seeberi*, has never actually been isolated from these sources, nor has the organism been successfully cultured. Because the organism has never been isolated from an aquatic reservoir, and has never been cultured, its taxonomic assignment has not been easy. In the past, the organism was presumed to be fungal, as it looks somewhat like a large yeast on histologic cross-section. Molecular analysis places the Rhinosporidium genus as a subclass within Class Choanozoa [85].

The geographic location with the highest incidence of rhinosporidiosis is southern India and Sri Lanka. In general, the disease can be found in any tropical region, and cases have been reported in the southeastern United States [85].

Rhinosporidiosis should not be confused with rhinoscleroma, a granulomatous lesion involving the nasal mucosa, caused by the bacteria, *Klebsiella rhinoscleromatis* (Gamma Proteobacteria, Chapter 7).

Infectious species:

Rhinosporidium seeberi (rhinosporidiosis)

Chapter 24

Archaeplastida

"If you want to make an apple pie from scratch, you must first create the universe."
Carl Sagan

Eukaryota
  Bikonta (2-flagella)
    Excavata
      Metamonada (Chapter 16)
      Discoba
        Euglenozoa (Chapter 17)
        Percolozoa (Chapter 18)
    Archaeplastida (Chapter 24)
      Viridiplantae
        Chlorophyta
          Trebouxiophyceae
            Chlorellales
              Chlorellaceae
                *Prototheca
    Chromalveolata (Chapters 19–21)
      Alveolata
        Apicomplexa (Chapter 19)
        Ciliophora (ciliates) (Chapter 20)
      Heterokontophyta (Chapter 21)
  Unikonta (1-flagellum)
    Amoebozoa (Chapter 22)
    Opisthokonta
      Choanozoa (Chapter 23)
      Animalia (Chapters 25–32)
      Fungi (Chapters 33–37)

There seems to be only one organism, in the entire kingdom of plants, that is capable of causing an infectious disease in humans. This organism is the algae, *Prototheca wickerhamii*. The emergence of plants, and the phylogenetic history of Genus Prototheca, makes a fascinating story.

J.J. Berman: Taxonomic Guide to Infectious Diseases. DOI: http://dx.doi.org/10.1016/B978-0-12-415895-5.00024-6

Archaeplastida is the superclass of the kingdom of plants. As the story goes, oxygenic photosynthesis was discovered by cyanobacteria, about 2.5 billion years ago. Subsequently, the cyanobacteria refined the process, and no organism other than cyanobacteria can be credited with oxygenic photosynthesis (though other organisms use photosynthetic processes that do not produce oxygen). All oxygenic photosynthesis in eukaryotes is accomplished with chloroplasts, symbiotic organelles descended from a captured cyanobacteria. An ancient member of Class Archaeplastida found that by engulfing a cyanobacteria, it could photosynthesize. This indentured relationship between Arachaeplastida and cyanobacteria became permanent, and every descendant of Class Arhaeplastida, including all the green plants, have benefited from their ancestor's theft of a cyanobacteria. It is generally believed that the acquisition of a cyanobacteria as a self-replicating synthesizing organelle, occurred only once in earth's history. Chloroplast-containing organisms other than those within Class Archaeplastida, seized chloroplasts (not cyanobacteria) from green algae or other members of Class Archaeplastida.

How do we know that chloroplast-containing organisms outside Class Archaeplastida acquired chloroplasts from other eukaryotic organisms, and not from cyanobacteria? By counting membranes around the chloroplast. Chloroplasts in the Archaeplastida are lined by two layers, corresponding to the inner and outer membranes of the original, indentured cyanobacteria. The chloroplasts of non-Archaeplastida eukaryotes have three or four membrane layers, suggesting that a member of Archaeplastida was engulfed, and the membranes of the chloroplast and the Archaeplastida were entrapped permanently in the host cell.

As aforementioned, the creation of a chloroplast, from a captured cyanobacteria, occurred once only, with all extant chloroplasts deriving from an early union between a member of Class Archaeplastida and a cyanobacterium. This assertion may be false, in one specific instance. *Paulinella chromatophora*, a member of the eukaryotic class Rhizaria, seems to have captured its own cyanobacteria and created its own permanent chloroplast-like organelle [86]. This conclusion is based, in part, on the dissimilarities between the photosynthesizing organelles of *Paulinella chromatophora*, and all other Eukaryotic chloroplasts.

Chloroplasts can be acquired as a temporary symbiont through a process called kleptoplasty. The kleptoplastic cell captures a chlorplast from an algae and uses the captured chloroplast for a short period (a few days to a few months) until the chloroplast degenerates. Fresh chloroplasts can be obtained by more of the same kleptoplastic behavior. Permanent chloroplasts are never found in Class Animalia. Thanks to kleptoplasty, one member of Class Animalia, the sacoglassan sea slug, has achieved a photosynthetic life style.

In the past, the algae were all classified as types of plants, primarily because they contained chloroplasts, and they looked more like plants than any other type of organism. Today, different types of algae are assigned to plant and non-plant eukaryotic classes. Three classes of algae are now assigned to Class Archaeplastida: Class Rhodophyta (red algae), Class Chlorophyta (green algae) and Class Glaucophyta [73]. The golden and brown algae are currently assigned to Class Heterokontophyta (Chapter 21), as are the diatomaceous algae. Because the term "algae" applies to organisms belonging to widely divergent biologic classes, it has lost much of its taxonomic relevance.

Archaeplastida
  Viridiplantae
    Chlorophyta
      Trebouxiophyceae
        Chlorellales
          Chlorellaceae
            *Prototheca

Although every member of Class Archaeplastida descended from a cell that contained a chloroplast, some of the descendants have lost the ability to photosynthesize. Genus Prototheca is a chloroplast-free subclass of Class Archaeplastida.

Prototheca species are unicellular algae ubiquitous in sewage and soil. Presumably, all humans are exposed to Prototheca sometime in their lives, but the number of human clinical infections is exceedingly rare. Protothecosis can occur in any of three clinical forms: cutaneous nodules, olecrenon (elbow) bursitis, and disseminated [87]. The two localized forms of protothecosis occur in either immune-competent or immune-compromised individuals, while the disseminated form seems to occur exclusively in immune-compromised individuals. Only about 100 cases were reported by 2004 [88].

Cutaneous protothecosis is characterized by nodules. Histologic examination usually shows chronic inflammation (lymphocytes and histiocytes dominating), and structures having the appearance of florets (i.e., resembling flowers). These structures are produced by the organism's thick wall, known as the theca (the "theca" of "prototheca"), within which are several autospores that formed by cleavage of the large organism [87]. Budding is not present.

Most human cases of protothecosis are caused by *Prototheca wickerhamii*. *Prototheca zopfii*, a species that causes disease in dogs, has been suspected to cause some cases of human disease.

Infectious species:

Prototheca wickerhamii (protothecosis)

Part IV

Animals

Chapter 25

Overview of Class Animalia

"If you think you are worth what you know, you are very wrong. Your knowledge today does not have much value beyond a couple of years. Your value is what you can learn and how easily you can adapt to the changes this profession brings so often."

Jose M. Aguilar

Here is the ancestral hierarchy leading to Class Animalia.

Eukaryota (nucleated cells)
  Unikonta (one flagellum)
    Opisthokonta (flagellum located at the posterior pole)
      Animalia (individual organisms develop from a blastocyst)

The highest class, Class Eukaryota, contains organisms with nucleated cells, and all the descendent subclasses will contain organisms that have nucleated cells. Furthermore, all classes that are not subclasses of Class Eukaryota, will not contain organisms that have nucleated cells. For example, Class Bacteria, which is not a subclass of Class Eukaryota, contains no organisms that have nuclei.

Class Unikonta, one of the two major subclasses of Class Eukaryota, is characterized by organisms that have a single flagellum. Class Opisthokonta is one of the subclasses of Class Unikonta. In Class Opisthokonta, the flagellum is located at the posterior pole. Class Animalia is a subclass of Class Opisthokonta. Animals have their own defining property (i.e., development from a blastula), along with inherited features descending from their ancestral superclasses. Hence, members of Class Animalia have nucleated cells, with one flagellum protruding from a posterior pole, as seen in sperm cells.

Animals are thought to have evolved from gallertoids; simple, spherical organisms suspended in ancient seas. The living sphere was lined by a single layer of cells enclosing a soft center in which fibrous cells floated in extracellular matrix. The earliest fossil remnants (seabed burrows and tracks) of these early animals date back to about 1 billion years ago. The classes of animals that we would recognize today (corresponding roughly to the chapter divisions of animals

J.J. Berman: Taxonomic Guide to Infectious Diseases. DOI: http://dx.doi.org/10.1016/B978-0-12-415895-5.00025-8

in this book) were all living during the Cambrian period, about 540 million years ago.

As the gallertoids evolved to extract food from the seabed floor, they flattened out. The modern animals most like the gallertoids are the placozoans, discovered in 1833, plastered against the wall of a seawater aquarium. These organisms are just under a millimeter in length and are composed of about 1000 epithelial cells. With the exception of being flat, rather than round, they resemble the gallertoids, with an outer lining of cuboidal cells, and an inner gelatinous matrix holding a suspension of fibrous cells.

What are the characteristic properties that set animals apart from non-animals? Animals are multicellular and form specialized tissues and organs. The most characteristic organ in animals is the gastrointestinal tract. A gastrointestinal tract is present in all eumetazoans (i.e., the true animals; see Glossary item, Metazoa). The gastrointestinal tract is only absent from the most primitive animals, Class Parazoa, comprising the placozoans and the porifera (sponges). The characteristic presence of a gastrointestinal tract explains why animals are sometimes referred to as traveling stomachs.

The single feature that taxonomists rely upon to distinguish animals from all other forms of life is the blastula. The blastula is an early embryo consisting of a sphere of cells enclosing a central cavity known as the blastocoel. Like the gastrointestinal tract, the blastula is present in all eumetazoan animals.

Three classes within Class Eukaryota (organisms with nucleated cells) are multicellular, with specialized cells and organs: classes Plantae, Fungi, and Animalia. These three classes account for virtually every organism that we can see with the naked eye. Consequently, prior to the invention of the microscope (about 1590 A.D.) and the advent of scientific observations of the microscopic world (about 1676 A.D.), these three classes accounted for the totality of the observed living world.

Fungi, plants, and animals can be distinguished by the method by which their included organisms develop. Fungi develop from spores. Plants, like animals, develop from embryos, but they do not have a blastula phase (i.e., the plant embryo does not develop a cavity). Animals develop from an embryonic blastula.

If you are a skeptic (and you should be), you may be wondering why the blastula serves as the defining feature that separates animals from other complex, multicellular organisms. For example, plants are complex organisms, with specialized tissues and organs. Plants, like animals, develop from embryos. Though it is easy to observe morphologic differences between plants and animals, it is difficult to imagine how the presence or absence of a blastula would account for the fundamental difference between plants and animals.

The importance of the blastula to Class Animalia may turn out to be something of a red herring. To form a blastula from a solid embryo, animals must have a fundamental property that is lacking in Class Fungi, Class Plantae, and in all eukaryote classes other than Class Animalia. This special property, unique to animals, is the ability to form specialized cell junctions. Specialized cell junctions account for the blastula stage of embryogenesis, and also accounts for the development of coelomic cavities, a digestive tract, and complex organs.

Unlike fungal cells (wrapped in chitin) and plant cells (wrapped in cellulose) animal cells have no structural molecule in cell walls to constrain them into any particular shape. With the exception of cells that produce intracellular bands of protein that force a spindle or elongated shape (such as actin, myosin, or desmin), most animal cells are relatively floppy bags of protoplasm, somewhat like water-filled balloons. If you press two water-filled balloons, of the same size, together, you will find that their surface of union is a flat circle. If you crowd together, onto a flat surface, a monolayer of floppy balloons, you will get a matrix of polygonal forms, wherein each straight edge of the polygon is bounded by an edge of an adjacent balloon. It is easy to see how a collection of spheres can, when pushed together, yield a cuboidal, or polygonal cell matrix, simulating an organ.

If polygonal epithelial cells were simply squeezed spheres, what would stop fluids from slipping through the spaces between the spheres? What would keep tissues from falling apart, when the squeezing stopped? The integrity of the polygonality of animal tissues is accomplished with two specialized structures that reside on the surface of animal cells: desmosomes and gap junctions. These structures, unique to Class Animalia, effectively zip together adjacent cells, creating a leak-proof continuum of epithelial cells. All animals contain epithelial cells that line the external surface of the animal (i.e., the skin), the gastrointestinal tract, and most of the internal organs (e.g. liver, pancreas, salivary glands).

When we look at histologic sections of any organ under a microscope, we typically see glands. Glands are round or tubular spaces lined by cuboidal or polygonal cells and filled with fluid or excreted proteins or mucus. When the cells that line glands are separated from one another with a digestive enzyme and suspended as single cells, in an aqueous solution, they appear to be spherical. Within glands, spherical cells transform into polygonal cells when they are pressed together and zipped up by specialized junctions.

The blastula phase of animal embryos is accomplished with the help of non-rigid cells connected by junctions. Fluids secreted by the cells accumulate in a central cavity, because the junctions are water-tight. An embryo, with a fluid-filled center, is known as a blastula. Only animals are equipped to produce a blastula. Presumably, the earliest animals, the gallertoids, were lined by non-rigid epithelial cells, zipped together with tight junctions. These junctions constrained the jelly-like fluid, produced by the epithelial cells, to the center of the gallertoid.

In eumetazoans (all animals other than sponges and placazoans), the gastrointestinal tract owes its existence to non-rigid epithelial cells with intercellular junctions [89]. Likewise, complex tissues and organs are built from non-rigid cells with intercellular junctions. Whereas the other complex, multicellular forms of life on earth (i.e., plants and fungi) opted for rigid cell walls, Class Animalia followed a different road. The fundamental biological feature that characterizes animals and distinguishes animals from all other organisms, can be traced to soft cells and specialized junctions.

Here is the schema for Class Eukaryota, including the subclasses of Class Animalia that constitute Chapters 25 to 32.

- Eukaryota
  - Bikonta (2-flagella)
    - Excavata
      - Metamonada (Chapter 16)
      - Discoba
        - Euglenozoa (Chapter 17)
        - Percolozoa (Chapter 18)
    - Archaeplastida (Chapter 24)
    - Chromalveolata (Chapters 19–21)
      - Alveolata
        - Apicomplexa (Chapter 19)
        - Ciliophora (ciliates) (Chapter 20)
      - Heterokontophyta (Chapter 21)
  - Unikonta (1-flagellum)
    - Amoebozoa (Chapter 22)
    - Opisthokonta
      - Choanozoa (Chapter 23)
      - Animalia (Chapters 25–32)
        - Eumetazoa
          - Bilateria
            - Deuterostomia
              - Chordata
                - Craniata (Chapter 32)
            - Protostomia
              - Ecdysozoa
                - Nematoda (Chapter 27)
                - Arthropoda
                  - Chelicerata (Chapter 29)
                  - Hexapoda (Chapter 30)
                  - Crustacea (Chapter 31)
              - Platyzoa
                - Platyhelminthes (Chapter 26)
                - Acanthocephala (Chapter 28)
      - Fungi (Chapters 33–37)

Chapter 26

Platyhelminthes (flatworms)

"The proof of evolution lies in those adaptations that arise from improbable foundations."

Stephen Jay Gould

Eukaryota
    Bikonta (2-flagella)
        Excavata
            Metamonada (Chapter 16)
            Discoba
                Euglenozoa (Chapter 17)
                Percolozoa (Chapter 18)
        Archaeplastida (Chapter 24)
        Chromalveolata (Chapters 19–21)
            Alveolata
                Apicomplexa (Chapter 19)
                Ciliophora (ciliates) (Chapter 20)
            Heterokontophyta (Chapter 21)
    Unikonta (1-flagellum)
        Amoebozoa (Chapter 22)
        Opisthokonta
            Choanozoa (Chapter 23)
            Animalia (Chapters 25–32)
                Eumetazoa
                    Bilateria
                        Deuterostomia
                            Chordata
                                Craniata (Chapter 32)
                        Protostomia
                            Ecdysozoa
                                Nematoda (Chapter 27)
                                Arthropoda
                                    Chelicerata (Chapter 29)

J.J. Berman: Taxonomic Guide to Infectious Diseases. DOI: http://dx.doi.org/10.1016/B978-0-12-415895-5.00026-X

- Hexapoda (Chapter 30)
- Crustacea (Chapter 31)
- Platyzoa
  - Platyhelminthes (Chapter 26)
    - Cestoda (tapeworms)
      - Cyclophyllidea
        - Taeniidae
          - *Echinococcus
          - *Taenia
        - Dipylidiidae
          - *Dipylidium
        - Hymenolepididae
          - *Hymenolepis
      - Pseudophyllidea
        - Diphyllobothriidae
          - *Diphyllobothrium
          - *Spirometra
    - Trematoda (flukes)
      - Digenea
        - Echinostomida
          - Fasciolidae
            - *Fasciola
            - *Fasciolopsis
        - Opisthorchiida
          - Heterophyidae
            - *Metagonimus
          - Opisthorchiidae
            - *Clonorchis
            - *Opisthorchis
        - Plagiorchiida
          - Dicrocoeliidae
            - *Dicrocoelium
          - Paragonimidae
            - *Paragonimus
        - Strigeatida
          - Schistosomatidae
            - *Schistosoma
            - *Trichobilharzia
            - *Schistosomatidae species
  - Acanthocephala (Chapter 28)
- Fungi (Chapters 33–37)

Every species of animal that infects humans is a descendant of Class Bilateria. Animals of Class Bilateria display bilateral symmetry (i.e., their bodies can be divided into two symmetrical halves by a plane that runs along a central axis). Class Bilateria includes all mammals and most large, complex animals. With one exception, members of Class Bilateria have a body cavity, known as a coelum. The exception is Class Platyhelminthes, the flatworms (from Greek "platys", flat, "helmis", worm). The prevalence of body cavities in all other members of Class Bilateria suggests body cavities are important. Humans have specialized body cavities in which organs are suspended. For example, the lungs are suspended in the pleural cavity, and the heart is suspended in the pericardial cavity. Lacking these cavities, members of Class Platyhelminthes lack lungs and a cardiovascular system. Oxygen is absorbed by simple diffusion; hence, platyhelminths are flat to maximize oxygen transfer.

Before going further, it is important to understand some of the taxonomic confusion that arises whenever the term "worm" is used to describe a human infection. Parasitic worms are called helminths. Class Platyhelminthes (Chapter 26) and Class Nematoda (Chapter 27) account for all the so-called helminthic diseases of humans.

Readers should be warned that the term "worm" has no taxonomic meaning; soft, squiggly organisms colloquially known as "worms" are scattered throughout animal taxonomy, with no close relationship to one another. A small squirming organism referred to as a "worm" may be an insect larva (i.e., not a helminth), or it may be one of several unrelated classes of organisms. Class Acanthocephala (Chapter 28) includes the thorny-headed worms. Class Annelida (earthworms) descends from Class Lophotrochozoa, which includes molluscs. Class Nematoda (roundworms) and Class Annelida (earthworms) are more closely related to spiders and clams, respectively, than either one is related to Class Platyhelminthes (flatworms). Many so-called worms are actually the larval forms of animals whose adult stage bears no resemblance to worms. An example is *Linguatula serrata* (Class Crustacea, Chapter 31), the agent causing tongue worm disease. The tongue worm is the larval stage of a crustacean. Likewise, the screw-worm (*Cochliomyia hominivorax*) is actually a type of fly. It is called a screw-worm because the disease is manifested by worm-like larvae growing in skin. The ineptly named "ringworm" infections are not caused by worms or by any animals; they are fungal infections of the skin. The word "worm" may even refer to a marsupial joey (Class Mammalia), which is typically a smooth hairless slug-shaped organism the size of a jellybean. When a word cannot be applied to a set of objects that are mutually related, what meaning does it convey? When you come across one of these forms of life, you either know its proper taxonomic class, or you do not. If you cannot classify the organism correctly, then no generic term will suffice.

Perhaps the most striking property of Class Platyhelminthes is its remarkable developmental plasticity. Class Turbellaria, a subclass of platyhelminths that contains no human pathogens, is adept at growing whole organisms from excised parts. A tiny fragment of planaria (*Schmidtea mediterranea*), several hundredths the size of the original organism, can regenerate into a full-sized planaria. The pathogenic species of Class Platyhelminthes, though lacking the full regenerative abilities of the planaria, have a great capacity to modify their life cycles to enhance their parasitism.

The flatworms that cause human disease fall into two subclasses: Class Cestoda (the tapeworms) and Class Trematoda (the flukes).

- Platyhelminthes
  - Cestoda (tapeworms)
    - Cyclophyllidea
      - Taeniidae
        - *Echinococcus
        - *Taenia
      - Dipylidiidae
        - *Dipylidium
      - Hymenolepididae
        - *Hymenolepis
    - Pseudophyllidea
      - Diphyllobothriidae
        - *Diphyllobothrium
        - *Spirometra

The generalized life cycle of cestodes applies to all members of the class and accounts for the pathologic manifestations of these infections.

Adult tapeworms live in the intestinal tract of an animal (the primary host), absorbing nutrients from partially digested food and other products in the bowel lumen. The adult tapeworm is composed of a scolex (the "head"), which attaches to the intestinal wall, followed by a neck, followed by the proglottids, that line up, one after the other, forming a "tape". The proglottids drop off from the end of the tape, and pass out of the organism in feces.

By the time the proglottid has dropped into the environment, it is gravid with infective eggs. The gravid proglottids are ingested by animals that eat grasses or food that has been contaminated by the proglottids. Animals infected by proglottids are the intermediate, or secondary, hosts. Eggs hatch out juvenile forms that migrate through the secondary host, eventually stopping to encyst in muscles and other organs. The cysts cause illness, in the secondary host, commensurate with their number, size, and anatomic locations.

Tapeworm infections for which humans are the primary host, are caused, in almost all cases, by eating undercooked tissues from an infected animal that is a secondary host. The disease that results consists of adult tapeworm disease, wherein the tapeworm, attached to the gut wall, lives off nutrients in the intestinal lumen.

Tapeworm infections for which humans are a secondary host are caused by eating food that is contaminated by proglottids and eggs dropped by a primary host. The disease that results consists of larval cysts growing in human tissues.

The six genera of cestodes that infect humans are: Hymenolepis, Echinococcus, Taenia, Dipylidium, and Diphyllobothrium and Spirometra.

Echinococcus species use humans as secondary hosts. This means that humans become infected when they ingest the eggs or the proglottids that were passed into the environment in the feces of primary hosts (usually dogs and other carnivores). The subsequent disease, echinococcosis, also known as hydatid disease (from Greek, "hydatid", meaning watery cyst), results from the growth of larval cysts (cysticercoids) within human organs. Because humans are seldom eaten by potential primary hosts, the human is a dead-end host for the organism. The severity of the disease is determined by the size and locations of the cysts. The most deadly forms of the disease produce large cysts in the central nervous system.

Several different species of Genus Echinococcus cause human disease: *Echinococcus granulosus*, *Echinococcus multilocularis*, *Echinococcus vogeli*, and *Echinococcus oligarthus*.

Taenia solium (the pork tapeworm) and *Taenia saginata* (the beef tapeworm) use humans as the primary host for adult tapeworms. The resulting disease is often referred to as taeniasis. Humans become infected when they eat undercooked meat from infected animals (pork for *Taenia solium*, and beef for *Taenia saginata*).

Taenia solium can also use humans as a secondary host. When humans ingest eggs or proglottids (from soil contaminated with the feces of infected pigs), the hatched larvae may produce cysts (cysticercoids) throughout the body. This disease is known as cysticercosis. When cysts occur in the brain, it is referred to as neurocysticercosis.

Thus, *Taenia solium* can cause two separate diseases in the human population: taeniasis when the human is a primary host; and cysticercosis when the human is the secondary host.

Dipylidium caninum uses humans as a primary host. Humans become infected when they accidentally ingest fleas or lice (the secondary hosts) whose tissues are infected with cysticercoids. The fleas and lice are carried by dogs. After ingestion by humans (usually children), the cysts develop into adult tapeworms in the intestine. One of the names given to this adult tapeworm infestation is "double-pore tapeworm disease". This name derives from the characteristic anatomic feature of Class Dipylidiidae wherein genital pores appear on both sides of proglottids.

Hymenolepis nana, the last-to-be-described human tapeworm of Class Cyclophyllidea, has a complex life cycle that can only be understood in terms of its ability to combine the primary and secondary phases of its life cycle within a single host organism.

Hymenolepis nana, the dwarf tapeworm (the Greek, "nana", means dwarf), occurs worldwide and is the most common cestode infection in humans. For *Hymenolepis nana*, humans serve as primary and secondary hosts, and both host roles occur concurrently in the same host. Here is how it works. Humans can become primary hosts when they eat animals (in this case, larval fleas and beetles) that contain infectious larval cysts (so-called cysticercoids). The larval cysts develop into adult tapeworms in the small intestine lumen, where gravid proglottids develop. Humans become secondary hosts of *Hymenolepis nana* when they ingest gravid proglottids or eggs dropped by a primary host. The hatched larvae penetrate into the mucosa of the intestinal villi, where they remain, to form larval cysts within the lymphatic channels of the intestinal submucosa.

In the case of *Hymenolepis nana*, the infectious cysts (of the secondary host) nestled in the intestinal submucosa can mature in-place, to produce adult organisms in the small intestine. Even stranger, eggs deposited in the small intestine, by the adult proglottids within the primary host, can encyst in-place, in the submucosa. This means that a secondary host can assume the role of a primary host, and a primary host can assume the role of a secondary host.

Because a single human can serve as both the primary and the secondary host of *Hymenolepis nana*, the net result is that humans infected with *Hymenolepis nana* may have chronic infections, characterized by a huge worm burden in the small intestine, composed of adult tapeworms and larval cysts, resulting in the fecal passage of enormous number of eggs, over a long period of time. Other tapeworms, including other species within Genus Hymenolepis, lack this life-cycle flexibility. It is not surprising that *Hymenolepis nana* is the most common tapeworm infection of humans, worldwide (while Ascaris lumbricoides, discussed in Chapter 27, is the most common helminth infection overall).

Rare cases of human hymenolepiasis are caused by the rodent tapeworms *Hymenolepis microstoma* and *Hymenolepis diminuta*.

Diphyllobothrium latum uses humans as a primary host. Humans become infested when they eat undercooked meat from an infected fish (the secondary host). *Diphyllobothrium latum* has an unusual way of passing larvae through several intermediate hosts before infecting the final intermediate host consumed by humans. Copepods become secondary hosts after ingesting eggs dropped into water along with the feces of an infected primary host (various mammals including humans). When copepods are eaten by fish, infectious larvae persist in the larval stage (the so-called plerocercoid larvae), using the fish as another intermediate host. Humans break the chain when they ingest plerocercoid larvae that hatch and become adult tapeworms in the human small intestine.

Spirometra species like *Diphylobothrium latum* belong to Class Pseudophyllidea, and their larvae can develop in stages, within two secondary hosts. In this case, humans are used as the second intermediate host, and are infected by the plerocercoid larvae (the same role held by fish in the life cycle of *Diphylobothrium latum*, see above). Humans are a dead-end host. The plerocercoid larvae migrate to subcutaneous tissues and various organs, producing an inflammatory response. The resulting diseases is called sparganosis. Clinical symptoms vary with the location of the larvae. Only a few hundred cases of sparganosis have been reported. The primary hosts vary and include dogs and birds. The first intermediate host is copepods. The natural second intermediate hosts are usually birds, reptiles, or amphibians. Humans usually contract the disease by consuming the raw flesh of one of the natural intermediate hosts or, more likely, drinking copepods, an animal small enough to go unnoticed in a glass of water.

Platyhelminthes
- Trematoda (flukes)
  - Digenea
    - Echinostomida
      - Fasciolidae
        - *Fasciola
        - *Fasciolopsis
    - Opisthorchiida
      - Heterophyidae
        - *Metagonimus
      - Opisthorchiidae
        - *Clonorchis
        - *Opisthorchis
    - Plagiorchiida
      - Dicrocoeliidae
        - *Dicrocoelium
      - Paragonimidae
        - *Paragonimus
    - Strigeatida
      - Schistosomatidae
        - *Schistosoma
        - *Trichobilharzia
        - *Schistosomatidae species

Trematodes (flukes) are shaped like flattened worms. All of the flukes that infect humans are members of Class Digenea. Members of Class Digenea, like all trematodes, have a sucker at one end (the mouth) and a second, ventral, sucker on its underside. A distinctive anatomic feature of species in Class Digenea is its outer coat, wherein the junctions between cells disappear, forming a cytoplasmic syncytium. The flukes in Class Digenea have a life cycle that requires a minimum of two hosts (from Latin "di", meaning two and "ginus", meaning type). In several cases, intermediate stages reproduce asexually. The complex life cycles of flukes, and their ability for larval stages to reproduce themselves, greatly expanding the number of larval organisms, are further indications of the remarkable generative abilities of Class Platyhelminthes (see above the planarian platyhelminth, *Schmidtea mediterranea*).

Here is the general life cycle of a Trematode:

1. Eggs are released into the environment (usually water) by adult trematodes within the primary host. For example, *Schistosoma haematobium* females release their eggs into the bladder lumen, where they are washed out with the urine stream.
2. The eggs hatch into swimming larvae known as miracidia, that penetrate their first intermediate host (a snail or other mollusc).
3. The miracidia, living in the first intermediate host, develop into a sac-like structure called a sporocyst.
4. The sporocyst, unlike cystic stages in tapeworms, is capable of self-reproduction (in some species), producing a daughter sporocyst. The sporocyst also produces the next stage of larval development, the redia, a larval form that has an oral sucker.
5. A redia, like the sporocyst, is capable of self-reproduction, and can produce more rediae. In addition, the redia is capable of producing a cercaria.
6. A cercaria is an organism that develops from germinal cells of a sporocyst or a redia. This means that the cercaria is not just a phase of larval development produced by morphologic transformation of one larval form into another; it is a new organism that arises from a particular type of cell within a larval organism. The cercaria is motile.
7. The motile cercaria may infect a new host (the primary host), where it becomes an adult fluke. Or the cercaria may transform into one of two dormant forms that persist in the environment, often attached to edible vegetation, or in another intermediate host. These two forms are: mesocercaria and metacercaria. The mesocercaria is a larval form of the organism while the metacercaria is an encysted form.

It is obvious that the flukes are a form of life with an incredibly complex and flexible life cycle. Trematode infections in humans can be simplified by remembering the following rule: All pathogenic classes of trematodes, with one exception, produce human infections wherein the human is the primary host (i.e., humans host the egg-laying adult fluke); and in which humans become infected by eating infectious metacercariae, that have settled on vegetation, or that have infected a second intermediate animal host. The exception to this rule is Class Schistosomatidae, which will be discussed at the end of the chapter.

Members of Class Fasciolidae produce adult flukes that live in the liver, gall bladder, and intestines of the primary host (e.g., humans). All pathogenic genera of Class Fasciolidae use freshwater snails as one of the intermediate hosts (though some species have expanded their host range to include other molluscans). Two polysyllabic diseases are caused by members of Class Fasciolidae: fascioliasis and fasciolopsiasis.

Most cases of fascioliasis are caused by *Fasciola gigantica* (liver fluke disease). This disease, common in Southeast Asia, occurs when a human ingests metacercaria that contaminate vegetables. The metacercaria develop in the small intestine, and eventually migrate to the biliary ducts of the liver. The adult flukes lay eggs that are passed in feces. A variety of animals serve as hosts for *Fasciola gigantica*, including cattle.

Fasciola hepatica, the sheep liver fluke, infects ruminants (particularly sheep) but may also infect humans. Human infection occurs when metacercaria on vegetation (particularly watercress) are ingested.

Fasciolopsiasis is caused by *Fasciolopsis buski*, a large (up to 7.5 cm length) fluke that lives in the intestines of the primary host (pigs and humans). The number of humans infected is about 10 million. Surprisingly, most infections are asymptomatic. Humans become infected by ingesting metacercaria on vegetation. The disease occurs almost exclusively in Southeast Asia and India.

Metagonimiasis is caused by any of several species of small flukes that live in the small intestines of the primary host. The metacercaria cysts attach to the underside of fish scales. Humans become infected when they eat undercooked fish that contain the metacercaria. The disease occurs wherever infected species of fish are eaten raw or uncooked. Most cases occur in Asia; cases are particularly common in Korea.

Class Opisthorchiidae contains two infectious genera: Clonorchis and Opisthorchis. Infectious genera within Class Opisthorchiidae produce the condition known as opisthorchiasis, sometimes imprecisely called clonorchiasis, in which adult flukes live in the bile ducts of the liver. Humans become infected after eating undercooked fish containing metacercaria.

Clonorchis sinensis, the Chinese liver fluke (also known as the Oriental liver fluke), infects about 30 million people. As the name suggests, most cases occur in Asia. Humans become infected when they eat uncooked or raw fish infected with the metacercaria. The metacercaria develop in the small intestine, and eventually migrate to the bile ducts. The adult flukes live in the bile ducts. These flukes feed on bile produced by the liver, excreted through the bile ducts and stored in the gall bladder. Hence, bile does not reach the small intestine; and the normal digestion of food, in the human host, is interrupted. Flukes in the liver produce a chronic inflammatory response, and untreated infections may eventually lead to the development of bile duct cancer (choledochocarcinoma and cholangiocarcinoma).

Opisthorchis felineus, the cat liver fluke, infects humans who eat undercooked fish containing the metacercaria. The adult fluke lives in the bile ducts of the liver. Untreated infections may lead to cirrhosis (diffuse fibrosis) of the liver and to an increased risk of liver cancer. Human infections of *Opisthorchis felineus* occur most often in Siberia.

Clonorchis viverrini, the Southeast Asian liver fluke, is likewise transmitted when humans ingest metacercaria in undercooked fish, and adult flukes live in the bile ducts.

Class Plagiorchiida contains two infectious genera: Dicrocoelium and Paragonimus.

Two species of Genus Dicrocoelium (*Dicrocoelium dendriticum* and *Dicrocoelium hospes*) are rare causes of liver fluke disease in humans. The fluke is small, narrow and long; hence, the alternate name, lancet liver fluke. The adult flukes live in the distal branches of the biliary tree, where they tend to produce mild disease (compared with the biliary infections produced by members of Class Opisthorchiidae, see above). The distinguishing feature of Dicrocoelium infections is the second intermediate host: the ant, *Formica fusca*. Humans become infected when they ingest ants infected by metacercariae; hence, the rarity of human disease.

Paragonimus westermani, along with dozens of less common species within Genus Paragonimus, causes the condition known as paragonimiasis. Paragonimus species are lung flukes. Adult flukes live in the respiratory tree of various infected animals (including rodents, pigs, and humans). Humans become infected when they eat undercooked crabs or crayfish that are infected with metacercariae. Adult flukes and their eggs produce pulmonary inflammation. Paragonimiasis is a common cause of hemoptysis (coughed blood) in endemic areas. Paragonimus eggs can be found in the sputum of infected individuals. About 22 million people are infected worldwide, with most cases occurring in Southeast Asia, Africa, and South America. Cases also occur in the United States.

Trematode infections by pathogenic members of Class Schistosomatidae are fundamentally different, in their method of transmission, from the previously discussed fluke infections. All the other subclasses of trematodes infect humans when the metacercaria (cysts) are eaten. The schistosomes infect humans when cercaria (the phase of larval development that precedes metacercaria), swimming in water, actively penetrate the skin of humans.

Schistosomiasis is the disease caused by any of the five pathogenic species of Genus Schistosoma: *Schistosoma haematobium* (urinary schistosomiasis), *Schistosoma intercalatum* (intestinal schistosomiasis), *Schistosoma japonicum* (schistosomiasis), *Schistosoma mansoni* (intestinal schistosomiasis), *Schistosoma mekongi* (Asian intestinal schistosomiasis). About 200 million people are infected by schistosomes. Infections are common in developing countries, with about half of the infections occurring in Africa.

Trichobilharzia regenti is another member of Class Schistosomatidae. Humans are an accidental dead-end host for this species. As in schistosomiasis, the cercaria penetrates the skin of humans, but the cercaria cannot develop further. The disease is a localized infection known as swimmer's itch or cercarial dermatitis. Other genera of Class Schistosomatidae are known to produce skin rashes in humans.

Infectious species:

Hymenolepis diminuta (rat tapeworm, rarely causing hymenolepiasis in humans)
Hymenolepis nana (hymenolepiasis)
Echinococcus granulosus (hydatid disease, echinococcal disease, cystic enhinococcosis, echinococcosis)
Echinococcus multilocularis (hydatid disease, echinococcal disease, alveolar enhinococcosis, echinococcosis)
Echinococcus vogeli (hydatid disease, echinococcal disease, polycystic echinococcosis, echinococcosis)
Echinococcus oligarthus (hydatid disease, echinococcal disease, polycystic echinococcosis, echinococcosis)
Taenia saginata (beef tapeworm disease)
Taenia solium (pork tapeworm disease, neurocysticercosis, cysticercosis)
Diphyllobothrium latum (diphyllobothriosis or broad tapeworm disease or fish tapeworm disease)
Dipylidium caninum (cucumber tapeworm disease, double-pore tapeworm disease)
Spirometra species (sparganosis)
Fasciola gigantica (fascioliasis, liver fluke disease)
Fasciola hepatica (fascioliasis, common liver fluke, sheep liver fluke)
Fasciolopsis buski (fasciolopsiasis)
Fasciolopsis magna (fasciolopsiasis)
Metagonimus yokagawai (metagonimiasis)
Metagonimus takashii (metagonimiasis)
Metagonimus miyatai (metagonimiasis)
Clonorchis sinensis, alternately *Opisthorchis sinensis* (Chinese liver fluke, oriental liver fluke)
Opisthorchis felineus (cat liver fluke)
Opisthorchis viverrini (southeast Asian liver fluke)
Dicrocoelium dendriticum (lancet liver fluke)
Dicrocoelium hospes (lancet liver fluke)
Paragonimus westermani (paragonomiasis, lung fluke)
Schistosoma haematobium (urinary schistosomiasis)
Schistosoma intercalatum (intestinal schistosomiasis)
Schistosoma japonicum (schistosomiasis)
Schistosoma mansoni (intestinal schistosomiasis)
Schistosoma mekongi (Asian intestinal schistosomiasis)
Trichobilharzia regenti (swimmer's itch)

Chapter 27

Nematoda (roundworms)

"Life is hard. Then you die. Then they throw dirt in your face. Then the worms eat you. Be grateful it happens in that order."

David Gerrold

- Eukaryota
  - Bikonta (2-flagella)
    - Excavata
      - Metamonada (Chapter 16)
      - Discoba
        - Euglenozoa (Chapter 17)
        - Percolozoa (Chapter 18)
    - Archaeplastida (Chapter 24)
    - Chromalveolata (Chapters 19–21)
      - Alveolata
        - Apicomplexa (Chapter 19)
        - Ciliophora (ciliates) (Chapter 20)
      - Heterokontophyta (Chapter 21)
  - Unikonta (1-flagellum)
    - Amoebozoa (Chapter 22)
    - Opisthokonta
      - Choanozoa (Chapter 23)
      - Animalia (Chapters 25–32)
        - Eumetazoa
          - Bilateria
            - Deuterostomia
              - Chordata
                - Craniata (Chapter 32)
            - Protostomia
              - Ecdysozoa
                - Nematoda, roundworms (Chapter 27)
                  - Secernentea
                    - Ascaridida

J.J. Berman: Taxonomic Guide to Infectious Diseases. DOI: http://dx.doi.org/10.1016/B978-0-12-415895-5.00027-1

    - Anisakidae
      - *Anisakis
      - *Pseudoterranova
      - *Contracaecum
    - Ascarididae
      - *Ascaris
      - *Baylisascaris
    - Toxocaridae
      - *Toxocara
    - Dioctophymatidae
      - *Dioctophyme
  - Spirurida
    - Spirurina
      - Filarioidea
        - Onchocercidae
          - *Brugia
          - *Loa
          - *Onchocerca
          - *Mansonella
          - *Wuchereria
    - Camallanida
      - Dracunculoidea
        - Dracunculidae
          - *Dracunculus
  - Oxyurida
    - Oxyuridae
      - *Enterobius
  - Rhabditida
    - Strongyloididae
      - *Strongyloides
    - Ancylostomatidae
      - *Ancylostoma
      - *Necator
    - Metastrongylidae
      - *Angiostrongylus
    - Trichostrongylidae
      - *Trichostrongylus
- Enoplea
  - Dorylaimida
    - Trichocephalida
      - Trichinellidae
        - *Trichinella
        - *Capillaria
      - Trichuridae

*Trichuris
Arthropoda
Chelicerata (Chapter 29)
Hexapoda (Chapter 30)
Crustacea (Chapter 31)
Platyzoa
Platyhelminthes (Chapter 26)
Acanthocephala (Chapter 28)
Fungi (Chapters 33–37)

Class Bilateria (animals with bilateral symmetry) contains two prominent subclasses: Class Deuterostomia and Class Protostomia. These two subclasses, like their parent class, have bilateral symmetry. Additionally, the deuterostomes and the protostomes have three germ layers (ectoderm, mesoderm, and endoderm).

In deuterostomes, a dent forms an anus; the mouth forms later. Humans and all vertebrates are deuterostomes.

Protostomes also form a dent, early in development, but subsequent events are somewhat controversial. In most cases, it seems, the primordial dent in protostomes produces an anus and a mouth.

Three large subclasses of animals comprise Class Protostomia: Class Platyzoa, Class Ecdysozoa, and Class Lophotrochozoa. Only the first two classes contain organisms that infect humans. Class Platyzoa contains Class Platyhelminthes (Chapter 26) and Class Acanthocephala (Chapter 28). Class Ecdysozoa accounts for Class Arthropoda (Chapters 29–31), and Class Nematoda (Chapter 27).

Nematodes, also known as roundworms, lack a circulatory system, and a respiratory system. They are all pseudocoelomates, meaning that their body cavities are not fully lined by mesoderm. Compare this with the flatworms (Class Platyhelminthes, Chapter 26), that are acoelomate (i.e., without body cavities). The Nematodes all have a tube-shaped digestive tract, open at both ends (mouth and anus). The presence of a complete digestive tract is another property that distinguishes the roundworms from the flatworms; the flatworm digestive tract has a single opening for the ingestion of food and the excretion of waste.

Nematodes are covered by a cuticle composed largely of extracellular collagen and other proteins (e.g. cuticulins) excreted by epidermal cells. The cuticle is shed repeatedly through various life stages of the nematode.

Two major subclasses of Class Nematoda contain the organisms that infect humans: Secernentea and Enoplea. These two classes have various anatomic features that distinguish one from the other. Their most relevant distinction, for healthcare workers, is that members of Class Secernentea are terrestrial dwellers, while members of Class Enoplea are marine inhabitants.

The many pathogenic members of Class Secernentea are best described within their subclasses.

Nematoda
Secernentea
Ascaridida
Anisakidae
*Anisakis
*Contracaecum
*Pseudoterranova

Anisakiasis is the disease produced by members of Class Anisakidae, which includes the following genera: Anisakis, Contracaecum, and Pseudoterranova. Humans are infected by eating undercooked shellfish infected with larvae. Most cases occur in countries where raw fish is regularly consumed. Only a handful of cases are reported in the USA annually. Humans are a dead-end host for the worms. After ingestion, the larvae travel to the small intestine, and live for a time, eventually dying without reproducing. In rare cases, they may cause abdominal obstruction, but their most clinically significant effect is through an acute allergic reaction. An acute enteritis often occurs, or, in some cases, a generalized anaphylactic reaction. Some fishermen are so highly sensitized that they develop hyper-immune reactions simply from handling fish or crustaceans that are infected with the larvae.

Nematoda
Secernentea
Ascaridida
Ascarididae
*Ascaris
*Baylisascaris

Ascaris lumbricoides, the cause of ascariasis, infects about 1.5 billion people worldwide, making it the most common helminth (worm) infection of humans [90]. Most cases occur in tropical regions, particularly in Africa and Asia. Humans are the exclusive primary host, for the organism.

Infection occurs when humans ingest embryonated eggs contaminating water or food.

The ingested embryonated eggs produce larvae in the small intestine. The larvae invade through the intestinal wall and into the portal system, where they are transported to the alveoli of lungs. From there, they invade into the bronchial system, move upwards to the pharynx, and drop back into the intestinal system via the esophagus. The worms reach adulthood in the intestine, where they can grow to nearly 40 cm in length and have a lifespan of up to two years. While in the lungs, they can produce a pneumonitis. *Ascaris lumbricoides* is not the only helminth that migrates through the lungs. In Class Rhabditida, *Ancylostoma duodenale*, *Necator americanus*, and *Strongyloides storcoralis* invade the pulmonary system.

Eggs produced by the adult worms leave the intestine via the feces and contaminate soil and water where sanitation is deficient. Adult worms may produce a wide range of symptoms due to obstruction of the intestines and of the ducts that connect with the intestine. Like members of Class Anisakidae, *Ascaris lumbricoides* may provoke an allergic reaction. An allergy to *Ascaris lumbricoides* may also precipitate allergic reactions to shrimp and dust mites, as these unrelated species share antigens in common with *Ascaris lumbricoides.*

Readers should be careful not to confuse ascariasis with the similar-sounding anisakiasis (see above).

Baylisascaris procyonis causes baylisascariasis in humans. Raccoons are the primary host. The adult nematode lives in the raccoon intestine and eggs are dropped with raccoon feces. The eggs of *Baylisascaris procyonis* can survive for years, and they are extremely resistant to disinfectants and heat. In rare circumstances, humans may become infected, as the secondary host, by ingesting eggs. Eggs develop into larvae in the human intestine, and the larvae migrate out of the intestines and through various organs, where they eventually encyst. Involvement of the central nervous system by encysted larvae is an extremely serious condition.

Nematoda
  Secernentea
    Ascaridida
      Toxocaridae
        *Toxocara

Humans are dead-end, accidental hosts for *Toxocara canis* (the dog toxocara) and *Toxocara cati* (the cat toxocara). Infected animals pass eggs in their feces. Larval development occurs within the eggs, and if matured eggs are ingested by humans, the larvae can hatch in the small intestine. The larvae migrate through various tissues: eyes, lung, liver, and brain being common destinations.

The disease caused by toxocara organisms is toxocariasis. During the migratory stage, the disease is often referred to as visceral larva migrans. Eye involvement is referred to as ocular larval migrans. Readers should not be lulled into a false sense of terminologic security. When toxocara migrate through the skin, the condition is NOT called cutaneous larva migrans: this term is reserved for cutaneous manifestations of *Ancylostoma brasiliense*. An immune response to the migrating toxocara larvae may produce eosinophilia (i.e., an increase of eosinophils in the peripheral blood), and the term eosinophilic pseudoleukemia has been used to describe this condition.

After a period of migration, the worms, which cannot mature further in the human body, encyst, to produce small, localized, permanent nodules in tissues.

Readers should not confuse "toxocara" with the similar-sounding "toxoplasma" (Class Apicomplexa, Chapter 19), a problem aggravated when insouciant clinicians use the abbreviated and ambiguous form "toxo," which can refer to either organism.

Nematoda
  Secernentea
    Ascaridida
      Dioctophymatidae
        *Dioctophyme

Dioctophyme renale, also known as the giant kidney worm, is a rare cause of human disease. Humans, one of many animals serving as the primary host, are infected by ingesting an undercooked second intermediate host (usually fish or frogs) that had, in turn, ingested the first intermediate host (a freshwater earthworm). Ingested larvae penetrate the human intestine and migrate to the liver. From the liver, they migrate to a kidney (usually unilateral, usually the right kidney), where they become adults. Eggs laid by the adult worm are excreted in the urine. The infestation of large adult worms typically leads to the destruction of the kidney, if left untreated. Human disease is rare, and can occur anywhere in the world.

Nematoda
  Secernentea
    Spirurida
      Spirurina
        Filarioidea
          Onchocercidae
            *Brugia
            *Loa
            *Onchocerca
            *Mansonella
            *Wuchereria

Filarial nematodes are string-like worms that are sufficiently small to fit into lymphatic vessels. About 150 million people are infected by the filarial nematodes (genera Brugia, Loa, Onchocerca, Mansonella, and Wuchereria) [91]. *Wuchereria bancroft* and *Brugia malayi*, together, infect about 120 million individuals [91]. Most cases occur in Africa and Asia.

The life cycle for these organisms involves a human primary host, in which the adult filarial worms produce juveniles (microfilaria) that migrate through lymphatics and blood vessels, where the microfilaria are sucked out by a secondary host (i.e., a blood-feeding mosquito or fly). In the secondary host, they develop into larvae. The secondary host serves double duty as a vector, by injecting larvae into the primary host when the insect has a blood meal. The larvae grow into adult threadworms within the primary host, a process that takes place over a year or more, and the cycle continues. The symptoms and severity of the disease are largely determined by the anatomic destinations of the migrating worms, and on the total load of worms, as multiple infections in the same person lead to a continuously increasing filarial burden.

Wuchereria bancrofti, *Brugia malayi*, and *Brugia timori* tend to cause lymph system obstruction, which can lead to elephantiasis. Elephantiasis is a condition wherein one or more extremities, usually the legs, are chronically swollen.

Other filarial worms preferentially inhabit the fatty tissue within the subcutis of skin: loa loa (the African eye worm), *Mansonella streptocerca*, and *Onchocerca volvulus*. The subcutaneous tissue is also the cause of infection by a nematode in Class Camallanida: *Dracunculus medinensis* (see below). *Mansonella perstans* and *Mansonella ozzardi* live in the abdominal peritoneum.

Onchocerca volvulus is the cause of river blindness, the second leading cause of infection-produced blindness (behind trachoma, caused by *Chlamydia trachomatis*, Class Chlamydiae, Chapter 13). The ocular pathogenicity of *Onchocerca volvulus* is caused by an endosymbiont, *Wolbachia pipientis* (see discussion in Class Alpha Proteobacteria, Chapter 5).

In addition to the filarial infections for which humans are the natural, primary host, there are reported cases of so-called zoonotic filariasis, in which humans are a dead-end host. The zoonotic infections, as in the filarial infections for which humans are the natural hosts, are all transmitted by blood-feeding insects. The filaria live for a time in human tissues, where they eventually die, producing a localized inflammatory reaction. Although various species of filaria are found in a wide assortment of animal hosts, including birds and reptiles, only mammalian hosts have been associated with zoonotic filariasis in humans [92].

Nematoda
    Secernentea
        Spirurida
            Camallanida
                Dracunculoidea
                    Dracunculidae
                        *Dracunculus

Dracunculus medinensis is the single pathogenic species in Class Dracunculidae. Dracunculiasis, also known as guinea worm, has a dramatic clinical presentation and treatment. The adult female worm bursts forth from the skin, usually just above or below the knee. The astute physician coaxes the living worm, onto a stick, which he or she then winds slowly, thus delivering the intact worm and relieving the patient of his parasitic burden. The Rod of Asclepius, historically symbolizing the practice of medicine, is inspired by the ancient ritual whereby the Dracunculus worm is extracted.

Humans become infected when they drink water that has been contaminated with copepods (tiny organisms in Class Crustacea, Chapter 31) containing a juvenile form of the organism. Readers will remember that copepods are a favored secondary host for cestodes (Class Platyhelminthes, Chapter 26). After ingestion by humans, the copepods die, and larvae are released, to penetrate the stomach wall. The male worms die, but the females migrate to subcutaneous tissue and continue to grow into adulthood (about one year later). The infectious cycle is perpetuated when the mature female adult (approaching a meter in length), pokes through the skin and releases its larvae, predestined for ingestion by secondary hosts (i.e., copepods).

As with several of the described nematodes (i.e., *Ascaris lubricoides*, *Enterobius vermicularis*, and the common filarial species) humans are the exclusive primary host of *Dracunculus medinensis*. Public health measures have drastically reduced the occurrences of dracunculiasis in Africa and in Pakistan.

Nematoda
  Secernentea
    Oxyurida
      Oxyuridae
        *Enterobius

Enterobius vermicularis, known as the pinworm in the USA, causes enterobiasis, sometimes called oxyuriasis.

Humans are the primary and exclusive host of the pinworm. Eggs are ingested, and the larvae emerge in the duodenum. The larvae migrate in the direction of peristalsis, as they mature into adults. The adults mate in the ileum and settle in the distal ileum, the proximal colon, or the appendix, where the females become bloated with eggs. Afterwards, the adult females (now about 1 cm in length) migrate to the anal skin where they lay their eggs. The eggs infect other humans or re-infect the original host, through fecal–oral contamination or through dispersal in the air (e.g., when bed linens are shaken).

The most common symptom of pinworm is anal itching. Cases have been reported of appendicitis caused by gravid pinworms obstructing the lumen of the appendix.

Readers should avoid confusion by the lay-terms for Enterobius infection. In the United States, *Enterobius vermicularis* is known as pinworm; in the UK, it is known as threadworm. Adding to the confusion, in the USA,

Strongyloides stercoralis is known as threadworm; in the UK, it is known as pinworm.

Nematoda
  Secernentea
    Rhabditida
      Strongyloididae
        *Strongyloides
      Ancylostomatidae
        *Ancylostoma
        *Necator
      Metastrongylidae
        *Angiostrongylus
      Trichostrongylidae
        *Trichostrongylus

Class Rhabditida is characterized by juvenile forms that are adept at free-living (usually in soil) and are endowed with an impressive skill-set; the ability to penetrate skin, to migrate through tissues, and to survive for extended periods.

Strongyloides stercoralis is a nematode that primarily infects humans; naturally occurring infection by *Strongyloidies stercoralis* in animals other than humans, is not known at this time. Humans become infected when larvae, passed from human feces, and free-living in contaminated soil, penetrate exposed skin and enter the circulation system. No secondary host is involved in the life cycle of *Strongyloides stercoralis*. Once in the bloodstream, *Strongyloides stercoralis* larvae move to the lungs, where they invade the bronchial tree, advance to the pharynx, and drop into the esophagus. When they reach the small intestine, they are adults, and female worms begin to lay eggs, thus renewing the infectious life cycle. Infections in humans are uncommon and occur in countries with poor sanitation.

Strongyloides has one important biological trick that contributes to its clinical presentation. Larvae that develop from eggs laid in the small intestine are capable of autoinfecting the host, by invading though the intestinal mucosa, or by invading through the anal skin (when they emerge from the large intestine). The larvae pass to the lung, renewing the infection cycle within their original human host. This step short-circuits the step wherein larvae dwell as free-living organisms in soil.

The consequences of auto-infection are several: infection can continue for a long time, sometimes extending throughout the lifespan of the host; the infectious load (i.e. number of organisms in the body) can be immense.

The term applied to a high infectious load of *Strongyloides stercoralis* is "hyperinfection syndrome" [93]. Immune-compromised patients are at highest risk for hyperinfection syndrome. When hyperinfection occurs, the disease can be fatal.

In addition to *Strongyloides stercoralis*, other species that have been reported to infect humans are: *Strongyloides fuelleborni* and *Strongyloides kellyi*.

Class Anyclostomadea contains the two species responsible for nearly all cases of hookworm disease in humans: *Ancylostoma duodenale* and *Necator americanus*. The two species have similar clinical presentations, but different geographic distributions. *Ancylostoma duodenale* is common in the Middle East, North Africa, and India. *Necator americanus* is common in North and South America, as well as parts of Africa and Asia. Hookworms infect about 600 million people.

Hookworm eggs and larvae live in soil. Like the previously described member of Class Rhabditida, *Strongyloides stercoralis*, larvae of *Ancylostoma duodenale* and *Necator americanus* penetrate the skin of human hosts. *Ancylostoma duodenale* may also infect by oral ingestion. The larvae invade tissues and travel to the lungs, where they travel up the bronchial tree, eventually dropping into the esophagus, passing down the alimentary tract to the small intestine, where they mature into adults. Eggs laid by the female adult worms pass into the environment, with feces. Humans are the only natural host for the hookworms. There is no secondary host.

The distinctive biological feature of the hookworms, distinguishing them from other nematodes, is hemophagia (i.e., blood eating). The hookworms suck blood from the host, producing anemia and malnutrition. Children are particularly vulnerable to the effects of hookworm infection, which may cause delays in mental and physical development.

Ancylostoma brasiliense is a hookworm of dogs and cats. It occasionally penetrates the skin of humans, a dead-end host, and causes localized inflammation of the skin or limited subcutaneous migration (cutaneous larva migrans or creeping eruption).

Class Metastrongylidae contains two genera that contain infectious organisms: Angiostrongylus and Trichostrongylus.

Angiostrongylus cantonensis is a parasitic nematode (roundworm) that causes angiostrongyliasis, the most common cause of eosinophilic meningitis in Southeast Asia. *Angiostrongylus cantonensis* lives in the pulmonary arteries of rats; hence its common name, rat lungworm. Snails are the most common intermediate hosts, where larvae develop until they are infective. Humans are incidental hosts of this roundworm, and may become infected through ingestion of raw or undercooked snails or from water or vegetables contaminated by snails or slugs or deposited larvae. Ingested larvae travel through the blood to the brain, where the larvae may die or may progress to juvenile adults before eventually dying. The dead and dying organisms produce an allergic inflammatory reaction in the brain, with an increase in eosinophils in cerebral spinal fluid. Other species of Genus Angiostrongylus that may infect humans include *Angiostrongylus costaricensis* and *Angiostrongylus mackerrasae*.

Genus Trichostrongylus contains a variety of species that infect a wide range of animals; ten species are known to infect humans. Humans are infected when larvae are ingested in contaminated water or vegetables. The larvae mature in the gut, and adult organisms live in the small intestine. Infections are often asymptomatic. Symptoms, when they occur, are typically those of enteritis. Eosinophilia (increased eosinophils in the blood) often occurs, and severe symptoms arise in some individuals.

Nematoda
  Enoplea
    Dorylaimida
      Trichocephalida
        Trichinellidae
          *Trichinella
          *Capillaria
        Trichuridae
          *Trichuris

The nematodes that cause infection in humans belong to the Class Secernentea or Class Enoplea. The human pathogens in Class Secernentea have been described (see above). Class Enoplea has three genera containing species that are infections in humans: Trichinella, Capillaria, and Trichuris.

Trinchinella spiralis is the cause of trichinosis, also known as trichinellosis. Humans are simultaneously the primary and the secondary host for this organism, as the adult worms live in the intestines, producing eggs that hatch into larvae within the body of the female adult. The female adult worms live for about six weeks. The larvae leave the adult, penetrate the intestine and migrate to muscles, where they wait to be eaten by another potential host. Larvae live within individual muscle cells. This means that the parasitic larva is much smaller than a single human cell. A full-length larva is about 80 microns in length. The larva curls tightly within the muscle cell, allowing it to fit in a tight space.

Humans typically become infected when they eat undercooked meat. More than 150 different animals have been reported as sylvatic hosts. Most infections in the USA come from eating undercooked pork. In the early decades of the twentieth century, hogs were fed on pig meat, thus magnifying the infectious burden in the hog population. Methods for cooking pork were lax, and trichinosis in humans was common. With improved, regulated diets for hogs, and with proper methods of cooking pork, the incidence of trichinosis in the United States has dropped. Clinically, trichinosis produces enteric symptoms when the adult worms are reproducing in the intestines; muscle aches when the larvae are invading muscle cells.

Hepatic capillariasis, caused by *Capillaria hepatica*, is a rare human infection, with only a few dozen cases reported, most occurring in children. Eggs in soil are ingested, hatched larvae penetrate the intestinal mucosa, the larvae are carried through the portal system to the capillaries of the liver, where they mature to adults. The adults lay their eggs in the liver. This leads to liver fibrosis. If the parasite burden is high, cirrhosis may eventually develop. The parasite infects a wide range of animals, but rats are the most likely source of human infections. When an infected rat dies, its body decomposes, and eggs within the liver are released to the soil, where the life cycle resumes.

Trichuris trichiura is called the human whipworm, named for its tightly wound, thick segment, from which a straight, thinner segment extends (i.e., handle and whip). The disease caused by *Trichuris trichiura* is trichiuriasis. Humans are the natural primary host for the organism. No secondary host is involved. Ingested eggs hatch in the small intestine, and larvae mature in the large intestine. The mature worms, which can live up to five years, anchor in the colonic mucosa and release eggs that pass out of the colon with feces. Eggs in the soil contaminate food, and the life cycle resumes.

The degree of morbidity is determined by the number of parasitic worms. Heavy infections can produce bloody diarrhea, anemia, and even rectal prolapse (from the aggregate weight of worms the rectum). Trichiuriasis often accompanies other parasitic infections, in the same individual.

Infectious species:

Anisakis simplex complex (anisakiasis)
Pseudoterranova decipiens (anisakiasis)
Ascaris lumbricoides (ascariasis, ascaris pneumonitis)
Baylisascaris procyonis (baylisascariasis, larva migrans with brain involvement)
Toxocara canis, the dog roundworm (toxocariasis, visceral larva migrans, ocularis larva migrans)
Toxocara cati, or *Tococara mystax*, or the feline roundworm (toxocariasis, visceral larva migrans, ocularis larva migrans)
Dioctophyme renale (giant kidney worm infection)
Enterobius vermicularis, also called pinworm in the USA and as threadworm in the United Kingdom, or sometimes as seatworm (enterobiasis or oxyuriasis)
Strongyloides stercoralis, known as threadworm in USA and pinworm in United Kingdom (strongyloidiasis)
Ancylostoma duodenale (hookworm, along with *Necator americanus*)
Necator americanus (hookworm, along with *Ancylostoma duodenale*)
Angiostrongylus cantonensis (angiostrongyliasis)
Angiostrongylus costaricensis (abdominal angiostrongyliasis, intestinal angiostrongyliasis)
Angiostrongylus mackerrasae (eosinophilic meningitis)
Trichostrongylus orientalis (trichostrongyliasis, trichostrongylosis)
Dracunculus medinensis (dracunculiasis, guinea worm disease)

Brugia malayi (lymphatic filariasis, elephantiasis)
Brugia pahangi (animal filariasis, rarely infecting humans) [94]
Brugia timori (lymphatic filariasis)
Loa loa (loa loa filariasis, loiasis, loaiasis, calabar swellings, fugitive swelling, tropical swelling, African eyeworm)
Mansonella ozzardi (serous cavity filariasis, seldom causes clinical disease, but generalized symptoms including fever, lymphadenopathy, joint pain, headache and pruritis are occasionally encountered)
Mansonella perstans, formerly *Dipetalonema perstans* (serous cavity filariasis, bung-eye disease)
Mansonella streptocerca, formerly *Dipetalonema streptocerca* (streptocerciasis)
Onchocerca volvulus (onchocerciasis, river blindness)
Wuchereria bancrofti (filariasis)
Trichinella nativa (trichinellosis)
Trichinella nelsoni (trichinellosis) [95]
Trichinella pseudospiralis (trichinellosis) [96]
Trichinella spiralis (trichinellosis, trichinosis)
Capillaria philippinensis (intestinal capillariasis)
Capillaria hepatica (hepatic capillariasis)
Trichuris trichiura, human whipworm (trichuriasis)

Chapter 28

Acanthocephala

"Nothing in biology makes sense except in the light of evolution."

Theodosius Dobzhansky

Eukaryota
- Bikonta (2-flagella)
  - Excavata
    - Metamonada (Chapter 16)
    - Discoba
      - Euglenozoa (Chapter 17)
      - Percolozoa (Chapter 18)
  - Archaeplastida (Chapter 24)
  - Chromalveolata (Chapters 19–21)
    - Alveolata
      - Apicomplexa (Chapter 19)
      - Ciliophora (ciliates) (Chapter 20)
    - Heterokontophyta (Chapter 21)
- Unikonta (1-flagellum)
  - Amoebozoa (Chapter 22)
  - Opisthokonta
    - Choanozoa (Chapter 23)
    - Animalia (Chapters 25–32)
      - Eumetazoa
        - Bilateria
          - Deuterostomia
            - Chordata
              - Craniata (Chapter 32)
          - Protostomia
            - Ecdysozoa
              - Nematoda (Chapter 27)
              - Arthropoda
                - Chelicerata (Chapter 29)
                - Hexapoda (Chapter 30)

J.J. Berman: Taxonomic Guide to Infectious Diseases. DOI: http://dx.doi.org/10.1016/B978-0-12-415895-5.00028-3

Crustacea (Chapter 31)
Platyzoa
Platyhelminthes (Chapter 26)
Acanthocephala (Chapter 28)
Archiacanthocephala
Moniliformida
Moniliformidae
*Moniliformis
Fungi (Chapters 33–37)

Acanthocephalans, also known as thorny-headed worms or spiny-headed worms, live purely parasitic lives. As is so often the case among dedicated parasites, these organisms have unburdened themselves of some anatomic features that may have been useful to their ancestors, but which serve no purpose when host resources are readily available. For taxonomists, this is a vexing problem; without inherited anatomic features, it is difficult to establish an organism's lineage, with any certainty.

Taxonomists have placed the Acanthocephalans as a subclass of Class Platyzoa, along with the Platyhelminthes (Chapter 26), and Class Rotifera, a class that does not contain human pathogens. Based on gene comparisons, it seems likely that Class Acanthocephala will soon be moved a notch over, to become a subclass of Class Rotifera.

All adult acanthocephalans have an extensible proboscis that is armed with hooks for attachment to gut wall. There is variation in the size and morphologic features of the acanthocephalans. In the case of the *Moniliformis moniliformis*, a known human pathogen, worms, found in stool, are just over one centimeter in length and about four millimeters wide. The proboscis is armed with 14 rows, each row with six to eight hooks.

There are well over 1000 species within Class Acanthocephala, infesting an enormous variety of animals, including dogs, squirrels, rats, birds, fish, and insects, including cockroaches. Like many dedicated parasitic animals, members of Class Acanthocephala display a bewildering array of life cycles, with one or more intermediate hosts. The complete life cycles of many of the known genera have not been fully established.

Parasitologists have taken a particular interest in Class Acanthocephala because these parasites have evolved a gruesome strategy for host parasitism known as "brain-jacking." All studied acanthocephalan species are able to alter the behavior of their intermediate hosts so as to increase the likelihood that the acanthocephalan will be delivered to its final host. In many cases, the parasite provokes the intermediate hosts to move to a vulnerable location (e.g., the surface of a pond, or a well-lit and exposed spot of soil), where the final host can eat them. The mechanism underlying brain-jacking is not fully understood, but it seems to be mediated, in at least some cases, by the parasite-induced release of host serotonin.

Acanthocephala
  Archiacanthocephala
    Moniliformida
      Moniliformidae
        *Moniliformis

Although all genera of Class Acanthocephala are parasitic in animals, none are particularly well adapted to life within humans. Human infection is extremely rare; when it occurs, it is secondary to the unintentional ingestion of whole or part of an uncooked, natural host. A single genus of Class Acanthocephala is known to have produced disease in humans: Moniliformis. Another genus, Apororhynchus is occasionally listed as a human parasite, but review of the literature yields no specific report documenting human pathogenicity. Although the literature describing acanthocephaliasis is scant, it seems that human infections are characterized by diarrhea, secondary to worm attachments to the walls of the small intestine. Reports of human infections usually come from Middle Eastern countries [97].

Infectious species:

Moniliformis dubius (Acanthocephaliasis)
Moniliformis moniliformis (Acanthocephaliasis)

Chapter 29

Chelicerata

"The truly privileged theories are not the ones referring to any particular scale of size or complexity, nor the ones situated at any particular level of the predictive hierarchy, but the ones that contain the deepest explanations."

David Deutsch

Eukaryota
  Bikonta (2-flagella)
    Excavata
      Metamonada (Chapter 16)
      Discoba
        Euglenozoa (Chapter 17)
        Percolozoa (Chapter 18)
    Archaeplastida (Chapter 24)
    Chromalveolata (Chapters 19–21)
      Alveolata
        Apicomplexa (Chapter 19)
        Ciliophora (ciliates) (Chapter 20)
      Heterokontophyta (Chapter 21)
  Unikonta (1-flagellum)
    Amoebozoa (Chapter 22)
    Opisthokonta
      Choanozoa (Chapter 23)
      Animalia (Chapters 25–32)
        Eumetazoa
          Bilateria
            Deuterostomia
              Chordata
                Craniata (Chapter 32)
            Protostomia
              Ecdysozoa
                Nematoda (Chapter 27)
                Arthropoda

J.J. Berman: Taxonomic Guide to Infectious Diseases. DOI: http://dx.doi.org/10.1016/B978-0-12-415895-5.00029-5

Chelicerata (Chapter 29)
  Arachnida
    Acari
      Sarcoptiformes
        Sarcoptidae
          *Sarcoptes
      Trombidiformes
        Demodicidae
          *Demodex
        Cheyletidae
          *Cheyletiella
Hexapoda (Chapter 30)
Crustacea (Chapter 31)
Platyzoa
  Platyhelminthes (Chapter 26)
  Acanthocephala (Chapter 28)
Fungi (Chapters 33–37)

The subclasses of Class Arthropoda that harbor human infectious organisms are Class Chelicerata, Class Hexapoda (Chapter 30), and Class Crustacea (Chapter 31). Class Chelicerata contains a variety of organisms that would appear, at first glance, to be unrelated: spiders, mites, ticks, scorpions, and horseshoe crabs. The common names do not help much. Horseshoe crabs are not true crabs; true crabs belong to Class Crustacea (Chapter 31).

All the members of Class Chelicerata have chelicerae, embryonic appendages that form prior to, and in the vicinity of, the mouth. In most species within Class Chelicerata, the chelicerae are feeding pincers. In spiders, the chelicerae are fangs. Another feature of the chelicerates, which helps to distinguish them from insects, is the absence of antennae.

The enormous difference in size between a horseshoe crab (60 cm or less) and a Demodex mite (less than 1 millimeter) is yet another reminder that biological classes are not determined by similarities (such as size), but by phylogenetic relationships (such as chelicerae). Aside from their class-specific chelicerae, chelicerates inherit the body plan of their superclass, Class Arthropoda. This means that they have a heart that pumps blood though a major body cavity called a hemocele.

Most members of Class Chelicerata are non-infectious in humans. Only three genera of Class Chelicerata live in, or on, humans, and both genera belong to the subclass of arachnids named Class Acari, which includes mites and ticks. Readers should not confuse mites and ticks with insects. Insects are members of Class Hexapoda (Chapter 30).

Ticks are vectors for a variety of infectious pathogens. Tick species *Ixodes scapularis* transmits *Babesia microti* (babesiosis), *Borrelia burgdorferi* (Lyme disease), and *Anaplasma phagocytophilum* (human granulocytic anaplasmosis) [78]. Tick-borne viruses include: Crimean-Congo hemorrhagic fever, tick-borne encephalitis, Powassan encephalitis, deer tick virus encephalitis, Omsk hemorrhagic fever, Kyasanur forest disease (Alkhurma virus), Langat virus, and Colorado tick fever.

In this book ticks are not considered infectious organisms. Basically, ticks leave the host after collecting their blood meal. Because ticks are temporary guests and do not actually live in humans, they will not be discussed further in this chapter.

Chelicerata
  Arachnida
    Acari (includes mites and ticks)
      Sarcoptiformes
        Sarcoptidae
          *Sarcoptes
      Trombidiformes
        Demodicidae
          *Demodex
        Cheyletidae
          *Cheyletiella

There are three infectious genera of mites: Demodex, Cheyletiella, and Sarcoptes. Genera Demodex and Cheyletiella belong to the Acari subclass Class Trombidiformes. Genus Sarcoptes belongs to the Acari subclass, Class Sarcoptiformes. Both of these classes contain extremely small mites that can live on humans without being identified with the naked eye.

Genus Sarcoptes contains one infectious organism for humans; *Sarcoptes scabiei*, the cause of scabies in humans and sarcoptic mange in animals. This small mite burrows into the superficial skin. Burrows are visible to astute searchers. Once in the skin, the mites produce an allergic reaction, producing itching. The name scabies comes from the Latin "scabere," to scratch. Scabies is an exceedingly common, global disease, with about 300 million new cases occurring annually. The organism infects humans through contact with other infected humans and animals. Spread of the scabietic rash is encouraged by scratching, which results in self-inoculation from one area of skin to another. The organism prefers intertriginous sites (skin creases such as groin, under breast, between fingers).

Scabies infestation can be particularly severe for immune-compromised patients, particularly those with AIDS. In this circumstance, the infection can become generalized, extending over the entire surface of the body, excluding only the face.

Demodex is a tiny mite that lives in facial skin. *Demodex folliculorum* favors the hair follicles. *Demodex brevis* favors the sebaceous glands. Both mites can be found in the majority of humans, and infections seldom produce clinical disease. It is common for pathologists to encounter these mites, on cross-sections of hair follicles or sebaceous glands, when studying histologic samples of facial skin; a finding of no consequence. In the rare instances when Demodex infections are heavy, a condition characterized by itching and inflammation may occur. This condition is known as demodicosis. When demodicosis occurs on the eyelashes, the condition is demodectic blepharitis. *Demodex canis*, an infection of dogs that can produce demodectic mange, may rarely cross-infect humans.

Cheyletiella mites infect the skin of dogs, cats, and rabbits. Occasionally, *Cheyletiella yasguri* or *Cheyletiella blakei* infect humans, producing a self-limited dermatitis, cheyletiellosis, one of several different diseases characterized by the occurrence of red, itchy bumps.

Infectious species:

Demodex folliculorum (demodicosis)
Demodex brevis (demodicosis)
Demodex canis (dog mite, rarely infecting humans; demodectic mange in dogs)
Cheyletiella yasguri (cheyletiellosis, cheyletiella dermatitis, walking dandruff)
Cheyletiella blakei (cheyletiellosis, cheyletiella dermatitis, walking dandruff)
Sarcoptes scabiei (scabies)

Chapter 30

Hexapoda

"It has become evident that the primary lesson of the study of evolution is that all evolution is coevolution: every organism is evolving in tandem with the organisms around it."

Kevin Kelly

Eukaryota
    Bikonta (2-flagella)
        Excavata
            Metamonada (Chapter 16)
            Discoba
                Euglenozoa (Chapter 17)
                Percolozoa (Chapter 18)
        Archaeplastida (Chapter 24)
        Chromalveolata (Chapters 19–21)
            Alveolata
                Apicomplexa (Chapter 19)
                Ciliophora (ciliates) (Chapter 20)
            Heterokontophyta (Chapter 21)
    Unikonta (1-flagellum)
        Amoebozoa (Chapter 22)
        Opisthokonta
            Choanozoa (Chapter 23)
            Animalia (Chapters 25–32)
                Eumetazoa
                    Bilateria
                        Deuterostomia
                            Chordata
                                Craniata (Chapter 32)
                        Protostomia
                            Ecdysozoa
                                Nematoda (Chapter 27)
                                Arthropoda

J.J. Berman: Taxonomic Guide to Infectious Diseases. DOI: http://dx.doi.org/10.1016/B978-0-12-415895-5.00030-1

- Chelicerata (Chapter 29)
- Hexapoda (Chapter 30)
  - Insecta
    - Hemiptera
      - Cimicoidea
        - Cimicidae
          - Cimicinae
            - *Cimex
    - Phthiraptera
      - Anoplura
        - Pediculidae
          - *Pediculus
        - Pthiridae
          - *Pthirus
    - Diptera
      - Calliphoridae
        - *Cochliomyia
      - Oestroidea
        - *Calliphoridae
        - *Sarcophagidae
        - *Dermatobia
    - Siphonaptera
      - Hectopsyllidae
        - *Tunga
- Crustacea (Chapter 31)

Platyzoa
- Platyhelminthes (Chapter 26)
- Acanthocephala (Chapter 28)

Fungi (Chapters 33–37)

Class Arthropoda accounts for 80% of the known species of animals and has three subclasses containing organisms that are pathogenic to humans: Class Chelicerata (Chapter 29), Class Hexapoda (Class 30), and Class Crustacea (Chapter 31).

Class Hexapoda is the largest class of animals, in terms of the number of different class species. Molecular evidence would suggest that Class Hexapoda first appeared about 425 million years ago, coinciding with the earliest fossils of large, vascular land plants [98]. Presumably, the species diversity of Class Hexapoda is intimately related with the diversity of land plants, particularly plants in Class Angiospermae, the flowering plants. Insects pollinate the plants, the plants serve as primary food sources for the insects; plant and insect co-evolve to serve one another's survival interests. As the name would suggest, members of Class Hexapoda have six ("hex") legs ("poda"). The insect body is built from fused segments, and the head has a mandible and a maxilla.

Class Hexapoda has two major subclasses: Entognatha and Ectognatha; hexapods with enclosed mouthparts and hexapods with exposed mouthparts, respectively. Class Insecta belongs to the Class Ectognatha (exposed mouthparts). Members of Class Entognatha contain some of the most common organisms on earth (e.g., Collembola, also known as springtails), but none infect humans [99]. All of the infectious members of Class Hexapoda belong to Class Insecta.

Despite its precise taxonomic definition, members of other animal classes are often confused with members of Class Hexapoda. This is particularly true for Class Chelicerata (Chapter 29), which contains spiders, mites, ticks, and scorpions.

Hexapoda
  Insecta
    Hemiptera
      Cimicoidea
        Cimicidae
          Cimicinae
            *Cimex

Class Hemiptera, a subclass of Class Hexapoda, are the so-called "true bugs." They are distinguished from other insects by their mouth parts, which are shaped as a proboscis and covered by a labial sheath. The mouth parts of Class Hemiptera are designed for sucking. Class Hemiptera includes cicadas and aphids. The triatome species that are vectors for *Trypanosoma cruzi* (Euglenozoa, Chapter 17) are members of Class Hemiptera.

It is worth reminding readers that the word "bug" has no taxonomic meaning, as it cannot be applied to any particular class of organisms. It is meaningless to use the word "bug" to refer to any small crawling hairless organism, as the organism may be an insect (Class Insecta), or a hexapod that is not an insect (Class Entognatha), or a mite (Class Chelicerata) or a pill bug (Class Crustacea).

Cimicidae are true bugs (i.e., Class Hemiptera). Bedbugs (*Cimex lenticularis*) suck blood and produce an inflammatory response, but they are not a true infection (they don't live in the skin), and they are not known to be carriers of other infectious agents. Bedbugs are included in this book because there is currently a bedbug epidemic in USA cities; bedbug enthusiasts would be offended if this species were omitted.

Hexapoda
  Phthiraptera
    Anoplura
      Pediculidae
        *Pediculus
      Pthiridae
        *Phthirus

Three species of lice (Class Pediculus and Class Phthirus) have adapted to infect humans: *Pediculus humanus capitus*, the head louse, *Pediculus humanus corporis*, the body louse, and *Phthirus pubis* (crab louse or pubic louse). These lice are all hematophagus (blood-sucking), and can produce an inflammatory skin reaction at the site of infection. The eggs of lice are called nits. The nits of head lice and pubic lice are deposited on hairs. The nits of body lice are deposited on skin and clothing.

Hexapoda
  Insecta
    Diptera
      Calliphoridae
        *Cochliomyia
      Oestroidea
        *Calliphoridae
        *Sarcophagidae
        *Dermatobia

Myiasis is an infection of the larvae (maggots) of flies (Class Diptera). The screw-worm (*Cochliomyia hominivorax*), despite its vivid name, is a type of fly.

Because insect larva are small and squishy, they are often mistaken for worms. The disease caused by *Cochliomyia hominivorax* is manifested by worm-like larvae growing in skin. Whereas myiasis caused by most other species of fly is an infection of dead animals, or necrotic parts of living animals (e.g., blowfly, flesh-fly), the screw-worm lays its larvae in living flesh (human or animal). Fortunately, the screw-worm has been eradicated in the USA Human and animal screw-worm infections occur in Central and South America.

Another fly that can infect living human skin is the human botfly (*Dermatoba hominis*). The botfly larvae attach to the skin, sometimes using a passive vector, such as a mosquito, to arrive at a convenient entry point, at which they burrow downwards. The botfly is found in Central and South America.

When the larvae of flies are accidentally ingested, they can be found anywhere in the intestinal tract. If they are actively growing, the condition is referred to as intestinal myiasis. If they are simply passing through, without larval growth, the condition is called intestinal pseudomyiasis. Both conditions are rare [100].

Hexapoda
  Insecta
    Siphonaptera
      Hectopsyllidae
        *Tunga

Fleas (Class Siphonaptera) are another family of insects that draw blood. Fleas carry a variety of infectious organisms. Fleas are vectors for bacteria: *Yersinia pestis* (plague), *Rickettsia typhi* (endemic typhus), *Rickettsia felis* (endemic typhus), *Bartonella henselae* (bacillary angiomatosis, infectious peliosis hepatis, cat-scratch disease, bartonellosis). Fleas are an intermediate host for *Hymenolepis diminuta*, a rare cause of human tapeworm disease (hymenolepiasis). Despite early speculation, fleas are not carriers of the HIV/AIDS virus.

With one exception, fleas do not live on or in humans. The exception is tungiasis, a skin disease caused by *Tunga penetrans* (alternately known as chigoe, jigger, and nigua), a very small flea. Female fleas burrow into the skin, producing intense localized inflammation and pain. The skin lesion is characterized by a black dot surrounded by edematous and erythematous (red) skin. The disease occurs in Africa, South America, and the Caribbean, where the infection rate can be very high (about 50% of the population in endemic areas such as Nigeria, Brazil, and Trinidad).

Infectious species:

Cimex lectularius, bedbug (bedbug bites)
Siphonaptera species (flea bites)
Pediculus humanus, head louse (pediculosis)
Pediculus humanus corporis, body louse (pediculosis)
Phthirus pubis (crab louse, pubic louse)
Cochliomyia hominivorax (screw-worm myiasis)
Calliphoridae species (blowfly myiasis)
Sarcophagidae species (flesh-fly myiasis)
Dermatobia hominis (human botfly myiasis)
Tunga penetrans (tungiasis, nigua)

Chapter 31

Crustacea

"Probably a crab would be filled with a sense of personal outrage if it could hear us class it without ado or apology as a crustacean, and thus dispose of it. 'I am no such thing,' it would say; 'I am MYSELF, MYSELF alone.'"

William James

- Eukaryota
  - Bikonta (2-flagella)
    - Excavata
      - Metamonada (Chapter 16)
      - Discoba
        - Euglenozoa (Chapter 17)
        - Percolozoa (Chapter 18)
    - Archaeplastida (Chapter 24)
    - Chromalveolata (Chapters 19–21)
      - Alveolata
        - Apicomplexa (Chapter 19)
        - Ciliophora (ciliates) (Chapter 20)
      - Heterokontophyta (Chapter 21)
  - Unikonta (1-flagellum)
    - Amoebozoa (Chapter 22)
    - Opisthokonta
      - Choanozoa (Chapter 23)
      - Animalia (Chapters 25–32)
        - Eumetazoa
          - Bilateria
            - Deuterostomia
              - Chordata
                - Craniata (Chapter 32)
            - Protostomia
              - Ecdysozoa
                - Nematoda (Chapter 27)
                - Arthropoda

J.J. Berman: Taxonomic Guide to Infectious Diseases. DOI: http://dx.doi.org/10.1016/B978-0-12-415895-5.00031-3

Chelicerata (Chapter 29)
Hexapoda (Chapter 30)
Crustacea (Chapter 31)
Maxillopoda
Pentastomida
Linguatulidae
*Linguatula
Porocephalidae
*Armillifer
*Porocephalus
Platyzoa
Platyhelminthes (Chapter 26)
Acanthocephala (Chapter 28)
Fungi (Chapters 33–37)

Class Arthropoda has three subclasses that contain organisms that are pathogenic to humans: Class Chelicerata (Chapter 29), Class Hexapoda (Class 30), and Class Crustacea (Chapter 31). Class Crustacea includes many of the menu items that are commonly called shellfish: lobsters, crabs, shrimp. It also contains smaller organisms, such as copepods, commonly found in plankton; and barnacles, which are found encrusted over ocean surfaces. Most crustaceans live in water, but some live on land. An example of the latter is organisms of Genus Armadillidium, variously known as woodlice, pill bugs, potato bugs, and other misleading names that would wrongly suggest a non-Crustacean identity. Crustaceans are characterized by a body plan composed of fused segments covered by a single carapace and protected by a hard exoskeleton. Molting occurs in all growing crustaceans.

Perhaps the most distinctive feature of Crustaceans is the multiple larval stages that precede the emergence of the adult form. These larval forms vastly amplify the complexity of carcinology (the study of crustaceans). In past years, many larval forms of crustaceans have been incorrectly identified as separate animal species.

Crustacea
Maxillopoda
Pentastomida
Linguatulidae
*Linguatula
Porocephalidae
*Armillifer
*Porocephalus

Only one subclass of Class Crustacea contains organisms that are infectious in humans: Class Pentastomida, the so-called tongue worms. This class contains three genera that cause different forms of the same disease: pentastomiasis [101].

The pentastome adult has the appearance of a worm. Unlike worms of Classes Platyhelminthes or Nematoda, tongue worms display the typical crustacean body type: fused segments all covered by a hard, chitinous carapace. Tongue worms have five anterior appendages, leading to the class name “Pentastomida”.

The primary hosts of tongue worms (i.e., the host that carries the adult form of the organism) are vertebrate animals. Typically, the adult tongue worm parasitizes the respiratory tract of these animals. Humans are the intermediate host, carrying only the larval forms. Transmitted by food, pentastomid eggs hatch in the intestines, and the larvae invade tissues to encyst anywhere in the body. The nasopharynx is the most likely place to find the cysts.

Typically, the cysts are walled off by an inflammatory reaction, and eventually become calcified. With the exception of the obstructive effect produced by the presence of a cyst, infections are often asymptomatic. In areas where pentastomiasis is common, the cysts are often discovered as incidental findings, at autopsy.

Pentastomiasis occurs most often in Africa, Malaysia, and the Middle East. Less often, the disease may occur in China or South America. In the Middle East, pentastomiasis is known as Halzoun.

Three genera of Class Pentastomida are involved: Linguatula, Armillifer, and Porocephalus. One of the more confusing terms associated with pentastomiasis is “porocephaliasis,” named for a pentastome genus, Porocephalus. The genus “Porocephalus” and the infection “porocephaliasis” should not be confused with “porocephaly,” a rare developmental disorder in which cysts or cavities are found in the brains of infants.

Infectious species:

Linguatula serrata (pentastomiasis, marrara, Halzoun syndrome, tongue worm disease, linguatulosis)
Armillifer armillatus (pentastomiasis, porocephaliasis)
Armillifer grandis (pentastomiasis, porocephaliasis)
Armillifer moniliformis (pentastomiasis, porocephaliasis)
Porocephalus crotali (pentastomiasis, porocephaliasis)

Chapter 32

Craniata

"The only interesting thing about vertebrates is the neural crest."

Thorogood

Eukaryota
  Bikonta (2-flagella)
    Excavata
      Metamonada (Chapter 16)
      Discoba
        Euglenozoa (Chapter 17)
        Percolozoa (Chapter 18)
    Archaeplastida (Chapter 24)
    Chromalveolata (Chapters 19–21)
      Alveolata
        Apicomplexa (Chapter 19)
        Ciliophora (ciliates) (Chapter 20)
      Heterokontophyta (Chapter 21)
  Unikonta (1-flagellum)
    Amoebozoa (Chapter 22)
    Opisthokonta
      Choanozoa (Chapter 23)
      Animalia (Chapters 25–32)
        Eumetazoa
          Bilateria
            Deuterostomia
              Chordata
                Craniata (Chapter 32)
                  Actinopterygii
                    Siluriformes
                      Loricarioidea
                        *Trichomycteridae
            Protostomia
              Ecdysozoa

J.J. Berman: Taxonomic Guide to Infectious Diseases. DOI: http://dx.doi.org/10.1016/B978-0-12-415895-5.00032-5

Nematoda (Chapter 27)
Arthropoda
Chelicerata (Chapter 29)
Hexapoda (Chapter 30)
Crustacea (Chapter 31)
Platyzoa
Platyhelminthes (Chapter 26)
Acanthocephala (Chapter 28)
Fungi (Chapters 33–37)

From the human point of view, Class Craniata is the most fascinating group of organisms on earth, because it includes ourselves. The most distinctive evolutionary characteristic of Class Craniata is the neural crest. In embryologic development, the neural crest derives from a specialized compartment of cells lying between the ectoderm and the primitive neural tube. The neural crest gives rise to the peripheral nervous system, to the connective tissue of the cranium, to several endocrine glands, and to the connective tissue component of the teeth. As a general rule, if an animal has something that we might recognize as a face, it is a member of Class Craniata.

Although members of Class Craniata contain some of the most intelligent and predatory species on the planet, virtually no members of the class are infectious in humans. Basically, the members of Class Craniata enjoy eating one another, but they seldom infect one another. In the case of human infections, there is one exception: *Trichomycteridae plectrochilus*, the toothpick catfish.

Craniata
Actinopterygii
Siluriformes
Loricarioidea
*Trichomycteridae

The toothpick catfish, also known as candiru, is a small fish, shaped like an eel, about an inch in length, that lives in the Amazon and Oronoco Rivers. The candiru normally infects riverine fish, inserting itself between the gills, attaching to tissue with the aid of barbs, and feeding off the host's blood (hematophagy). There are a few case reports, and many anecdotal accounts, of candiru entering the urethra, the vagina, or the anus, of swimming or wading humans. Removal of the fish may require surgical excision.

Infectious species:

Trichomycteridae plectrochilus, candiru, toothpick fish, parasitic catfish of the Amazon (candiru disease)

Part V

Fungi

Chapter 33

Overview of Class Fungi

Definition of a lexicographer: "harmless drudge."

Samuel Johnson

Eukaryota
  Bikonta, 2-flagella (Chapter 16–21)
  Unikonta, 1-flagellum
    Amoebozoa (Chapter 22)
    Opisthokonta
      Choanozoa (Chapter 23)
      Animalia (Chapters 25–32)
      Fungi
        Zygomycota (Chapter 34)
        Dikarya
          Basidiomycota (Chapter 35)
          Ascomycota (Chapter 36)
        Microsporidia (Chapter 37)

In the past decade, Class Fungi has become the most intellectually frustrating branch of clinical microbiology. There are many reasons why mycology (the study of fungi) has become so very difficult.

1. Number of offending organisms. Approximately 54 fungi account for the vast majority of fungal infections, the actual number of fungi that are pathogenic in humans is much higher. To provide some idea of the ubiquitous nature of fungi, it is estimated that, on average, humans inhale about 40 conidia (spores from Class Ascomycota) each hour. Most of these organisms are non-pathogenic under normal circumstances. However, in the case of immune-compromised patients, or in the case of patients who provide a specific opportunity for ambient fungi to attach and grow within a body (e.g. an indwelling vascular line), an otherwise harmless fungus may produce a life-threatening illness. As the number of immune-compromised patients increases, due to transplants, AIDS, cancer treatment, long-term steroid use; and with the proliferation of medical

J.J. Berman: Taxonomic Guide to Infectious Diseases. DOI: http://dx.doi.org/10.1016/B978-0-12-415895-5.00033-7

devices that provide potential entry points for fungi, the number of newly recognized fungal pathogens will increase. It is estimated that there are about 20 new fungal diseases reported each year [102]. If the number of diseases caused by other types of organisms (i.e. bacteria, protists, animals, viruses and prions) remains steady, then it will not be long before the number of different fungal diseases exceeds the number of different diseases produced by all other organisms, combined.

2. Increased sensitivity of diagnostic tests. It is now possible to identify heretofore undiagnosed cases of pathogenic species [103]. In the past, when clinical mycology laboratories had fewer of the sophisticated tests available today, it was easier to lump diseases under a commonly encountered species or a genus. For example, *Aspergillus fumigatus* is a common cause of severe pulmonary infections in immune-compromised patients. With advanced typing techniques, an additional 34 species of Aspergillus have been isolated from clinical specimens [102].
3. Unstable taxonomy. Class Fungi has recently undergone profound changes, with the exclusion of myxomycetes (slime molds) and oomycetes (water molds), and the acquisition of Class Microsporidia (Chapter 37). The instability of fungal taxonomy impacts negatively on the practice of clinical mycology. When the name of a fungus changes, so must the name of the associated disease. Consider "*Allescheria boydii*"; people infected with this organism were said to suffer from the disease known as allescheriasis. When the organism's name was changed to *Petriellidium boydii*, the disease name was changed to petriellidosis. When the fungal name was changed, once more, to *Pseudallescheria boydii*, the disease name was changed to pseudallescheriasis [102]. All three names appear in the literature (past and present).
4. Classification by morphologic features of reproduction. Unfortunately, for the taxonomist, fungal organisms have two options for reproduction: sexual and asexual. Both these forms of reproduction have their own morphologic appearances, in the same species of organism. Factors that determine the mode of reproduction for a cultured fungus are a mystery. It is possible for one mycologist to observe a fungus that reproduces exclusively asexually, while another mycologist, observing the same species in a culture dish or growing in the wild, may observe sexual reproduction (e.g. fruiting bodies). Depending on the phase of reproduction observed, and ignorant of the existence of an alternate morphologic form, taxonomists have assigned different names (sexual and asexual) to the same organism. Rather than harmonizing a dichotomous nomenclature under one preferred name, the ICBN has ruled that it is acceptable to assign two different binomials to an organism: a sexual (also called teleomorphic, perfect, or meiotic form), and an asexual (also called anamorphic, imperfect, or mitotic form). For instance, two binomials legitimately apply to the same organism: *Filobasidiella neoformans* (the teleomorphic

form) and *Cryptococcus neoformans* (the anamorphic form). Clinical mycologists prefer the asexual name, because it is the asexual form that grows in human tissues.

5. Unclassifiable organisms. Many fungi have never exhibited sexual reproduction in culture. Many other fungi cannot be cultured. A special pseudo-class of fungi, deuteromycetes (spelled with a lowercase "d," signifying its questionable validity as a true biologic class) has been created to hold these indeterminate organisms until definitive classes can be assigned. At present, there are several thousand such fungi, sitting in a taxonomic limbo, until they can be placed into a definitive taxonomic class [102].
6. Variable morphologies. Pathogenic fungi grow within human tissues without reproduction (asexual or sexual). They typically grow in tissues as an expanding colony of hyphae or yeasts (so-called vegetative growth). The pathologist who observes fungal infections in human tissues typically reaches a diagnosis on clinical features and the somewhat restricted morphologic features of the fungus in biopsied tissue. Adding to the general confusion, fungal specimens grown in culture may have a different morphology from that of the same fungus growing in human tissue. This situation is very different from that of bacterial infections, which have the same morphology in tissues as they have in the culture dish.
7. An historical blunder. Class fungi got off to a very bad start when classical taxonomists mistook these organisms for plants. To this day, academic mycologists are employed by Botany Departments, and fungal taxonomy is subsumed under the International Code of Botanical Nomenclature (ICBN) [102]. Superficially, fungi resemble plants; both classes of organisms have members that emerge from the ground. We now know that fungi descended from a flagellated organism. Class Fungi was eventually reassigned to Class Opisthokonta (unikonts with a posterior flagellum), making Class Fungi a sister class to Class Animalia. A misunderstanding, based on an incorrect assumption related to the absence of a defining feature (in this case a flagellum) led to one of the most jarring re-assignments in the classification of living organisms (see Glossary item, Negative classifier).

Keeping these seven points in mind, the members of Class Fungi have a common ancestral lineage, share a set of common biological properties, and can be divided into four distinctive subclasses: Class Zygomycota (Chapter 34), Class Basidiomycota (Chapter 35), Class Ascomycota (Chapter 36), and Class Microsporidia (Chapter 37).

It is ironic that the important clue to the phylogenetic origin of fungi rests on the presence of a posterior flagellum, as fungi, with only one exception, lack flagella altogether. The only fungal class with a flagellum is Class Chytrid, considered to be the most primitive fungal class [73]. The chytrids are not listed in the schema of pathogenic organisms (below) because there are no known chytrids that cause human infection. However, chytrids, a rare aquatic fungus (most fungi grow in soil), are currently ravaging amphibian populations. The chytrid *Batrachochytrium dendrobatidis* seems capable of infecting thousands of different amphibian species and threatens many of these species with extinction.

Non-aquatic fungi (i.e. all classes of fungi other than the chytrids) lack flagella, presumably lost when these fungi adjusted to life in soil, and no longer needed a flagellum to propel themselves through water. For most of the fungi that are pathogenic in humans, individual fungal organisms grow on a surface or as a mycelial mass in the soil. Propagation occurs when asexual or sexual spores are expelled into the air and wafted to another location.

Aside from the ancestral single posterior flagellum, that establishes a close relationship between fungi and animals, there is also the presence of chitin. Chitin is a long-chain polymer built from units of N-acetylglucosamine, and found in the cell walls of every fungus. It is analogous to cellulose, which is built from units of glucose. Importantly, chitin is never found in plants, and cellulose is never found in fungi. Aside from its presence in fungi, chitin is found in some member of Class Protoctista and in some members of Class Animalia (particularly arthropods). Chitin is the primary constituent of the exoskeleton of insects. The important structural role of chitin in fungi and animals should have been a clue to the close relationship between these two classes. It happens that chitin was not discovered until 1930 (by Albert Hoffmann); well before that time, Class Fungi had been incorrectly assigned to the plant kingdom.

Lastly, fungi and animals are heterotrophic, acquiring energy by metabolizing organic compounds obtained from the environment. Plants, unlike animals and fungi, are phototropic autotrophs, producing organic compounds from light, water, and carbon dioxide.

Interactions between fungi and humans vary, following one or more of the following scenarios, listed in order of increasing clinical consequence:

1. The fungus grows in the external environment, usually in soil or on plants, never interacting in any way with humans.
2. Spores and asexual reproductive forms are emitted into the air. In warm and tropical locations, fungal elements are the predominant particulate matter found in air samples. Humans are exposed constantly to a wide variety of fungi just by breathing (spores and conidia), by ingestion (fungi grow on the plants we eat), and by direct skin contact with fungal colonies in soil and airborne organisms.
3. After exposure, fungi may leave, without colonizing (e.g. you inhale them, and then you exhale them, and they're gone).
4. After exposure, fungi may transiently colonize a mucosal surface, such as the oral cavity, the nose, the gastrointestinal tract, the respiratory

tract, or the skin. Once on a mucosal surface, an acute allergic response may occur (e.g. sneezing). After a time, the colony fails to thrive due to an inhospitable environment (e.g. insufficient food, poor ionic milieu, effective host immune response).

5. After exposure, fungi permanently colonize the mucosal surface, with no clinical effect. Candida species commonly colonize the mouth and the vagina. Aspergillus species may colonize the respiratory surfaces (e.g. bronchi). In many cases, we simply carry fungal colonies as commensals (organisms that live within us, without causing disease).
6. Colonies persist, but the host reacts with an acute or chronic immune response. Chronic allergic aspergillosis of the bronchi is a good example. The patient may have a chronic cough. Microscopic examination of bronchial mucosa may reveal some inflammation, the presence of eosinophils, and the occasional hypha. Sometimes the host response is granulomatous, producing small nodules lining the bronchi, containing histiocytes and lymphocytes. A truce between the fungal colony and the host response is sometimes attained, in which the fungus colonies never leave, the inflammation never regresses, but the fungus does not invade into the underlying mucosa.
7. Fungi invade through the mucosa into the submucosa and underlying tissue. These locally invasive infections often manifest as a fungal ball, consisting of varying amounts of inflammatory tissue, necrosis, and fungal elements.
8. Fungal elements invade into lymphatics, traveling with the lymph fluid, and producing regional invasive fungal disease along the route of lymphatic drainage. The prototypical example of this process is found in infections with *Sporothrix schenckii*, which typically gains entrance to the skin, from the soil, through abrasions. Infection yields multiple skin papules, emanating from the point of primary infection (usually the hand or the foot), and following line of lymphatic drainage.
9. Fungal elements invade into blood vessels.
10. Fungal elements become a blood constituent (i.e. fungemia) and disseminate throughout the body.
11. Fungal elements spread throughout the body to produce invasive fungal infections in multiple organs.

The most perplexing aspect of fungal infections is that a single fungus may manifest itself by any and all of these biologic options (e.g. Aspergillus and and Malassezia species, see below). In general, the more immune-competent the individual, the less likely that a fungal infection will become clinically significant or life-threatening.

Readers should be aware that pathologists have developed a wide variety of techniques to identify fungi based on their morphologic features in tissue biopsies (e.g. the presence or absence of pigment, the presence or absence of hyphal septation, the presence or absence of hyphal branching, the angulation of branches, hyphal thickness, the presence or absence of yeast forms, whether yeast forms grow by budding, the morphologic appearance of buds, etc.). Pathologists also use anatomic information to help identify fungal species (e.g. whether the infection is superficial or deep, local or systemic, the anatomic site of the lesion), and clinical history (whether the infection occurs in an immune-deficient patient, a diabetic, a child, a gardener, etc.).

Clinical mycologists find it useful to determine whether a fungus is thermally dimorphic. Thermally dimorphic fungi grow in the laboratory as yeasts (round organisms) or as hyphae, depending on temperature. For example, *Penicillium marnefei* grows as hyphae at room temperature and as yeasts at body temperature. *Penicillium marnefei* happens to be a non-pathogenic fungus, but it is an empiric observation that most of the highly pathogenic fungi capable of causing disease in healthy individuals, are thermally dimorphic. These organisms are: *Coccidioides immitis*, *Paracoccidioides brasiliensis*, *Blastomyces dermatitidis*, and *Histoplasma capsulatum. Candida albicans* and *Sporothrix schenckii* are also thermally dimorphic and infect immune-competent individuals, but typically cause mild, localized disease. All of these dimorphic fungi grow in human tissue as yeast forms, and all happen to be members of Class Ascomycota. Here's the problem; thermal dimorphism is a property that is sometimes clinically useful, but sometimes not, as it includes both harmful and harmless fungi. Thermal dimorphism is applicable only to fungi that can be cultured (for example Lacazia loboi, see below, cannot be cultured and dimorphism cannot be determined). Thermal dimorphism is sometimes taxonomically specific and sometimes not; most of the pathogenic dimorphic fungi belong to Class Ascomycota, but some do not (e.g. *Ustilago maydis*, the dimorphic fungus that produces smut in corn, is a member of Class Basidiomycota). Thermal dimorphism has value to clinical mycologists, who are well versed in the limitations of its clinical utility. For the rest of us, non-taxonomic approaches to medical mycology are devices best left to the experts.

With the aforementioned caveats, there are a few general properties of the fungi that most students will find useful.

1. Fungi propagate by ejecting reproductive elements into the air. Humans become infected when they inhale, ingest, or come into surface contact with fungi that contaminate air, soil, and water. Animal vectors are not required, and students need not memorize long lists of fungal vectors. Animals can, however, serve as reservoirs for fungi (e.g. Microsporidia, Chapter 37). If a fungus is growing in your environment, the overwhelming likelihood is that you are constantly exposed to numerous potentially infective fungal elements.
2. Most pathogenic fungi are globally ubiquitous, or they reside in the tropics. Thus, we seldom need to memorize their geographic distribution. The

few exceptions (e.g. Histoplasma and Coccidioides) are mnemonically tolerable.

3. Very few fungal diseases are contagious from person to person, though there are exceptions (e.g. tinea infections).
4. Fungal colonization, that does not result in disease, is quite common. Pneumocystis, Aspergillus, and Cryptococcus are just a few examples of potentially life-threatening infections that are found to inhabit the lungs of a significant percentage of healthy persons. When colonized individuals become immunosuppressed, endogenized fungi may emerge as serious pathogens.
5. Though there are over a million fungal species, and hundreds of potential fungal pathogens, the vast majority of human fungal diseases can be accounted for by a few dozen genera, falling into four classes: Class Zygomycota (Chapter 34), Class Basidiomycota (Chapter 35), Class Ascomycota (Chapter 36), and Class Microsporidia (Chapter 37). By far, Class Ascomycota contains the majority of the pathogenic fungal organisms, with 20 infectious genera. Readers are highly encouraged to memorize the fungi that fall into the classes with fewer infectious organisms (Zygomycota, Basidiomycota, and Microsporidia); all the other fungal pathogens will belong to Class Ascomycota.
6. Most fungal diseases do not occur in immune-competent individuals. Of the hundreds of fungal infections that can occur in humans, only a dozen or so produce disease in healthy persons. With few exceptions (e.g. Cryptococcus Gattii, Class Basidiomycota, Chapter 35), the clinically serious systemic mycoses that regularly occur in otherwise healthy individuals belong to Class Ascomycota.

Chapter 34

Zygomycota

"You can't teach an old dogma new tricks."

Dorothy Parker

Eukaryota
  Bikonta, 2-flagella (Chapter 16–21)
  Unikonta, 1-flagellum
    Amoebozoa (Chapter 22)
    Opisthokonta
      Choanozoa (Chapter 23)
      Animalia (Chapters 25–32)
      Fungi
        Zygomycota (Chapter 34)
          Mucormycotina
            Mucorales
              Mucoraceae
                *Rhizopus
                *Mucor
                *Absidia
              Syncephalastraceae
                *Syncephalastrum
          Entomophthoramycotina
            Entomophthorales
              Basidiobolaceae
                *Basidiobolus
              Ancylistaceae
                *Conidiobolus
        Dikarya
          Basidiomycota (Chapter 35)
            Agaricomycotina
          Ascomycota (Chapter 36)
            Pezizomycotina
            Saccharomycotina

J.J. Berman: Taxonomic Guide to Infectious Diseases. DOI: http://dx.doi.org/10.1016/B978-0-12-415895-5.00034-9

Taphrinomycotina
Microsporidia (Chapter 37)

There are four classes of fungi that contain pathogenic organisms: Class Zygomycota (Chapter 34), Class Basidiomycota (Chapter 35), Class Ascomycota (Chapter 36), and Class Microsporidia (Chapter 37).

Like all the major divisions of fungi, Class Zygomycota is characterized by its sexually reproductive form. In the zygomycotes, the sexual form is the zygospore. As with much of clinical mycology, the defining morphologic features of the zygomycotes are never observed in clinical specimens. In tissues, these organisms are present as hyphal forms, without a yeast phase. In the laboratory culture dish, they are present as hyphal colonies with asexual reproductive forms (sporangia containing spores, and free spores). The identification of species is typically made by expert evaluation of the available structures: hyphae and sporangia. In many instances, members of Class Zygomycota can be distinguished from members of the other major classes of fungi (i.e. ascomycotes and basidiomycotes) by three features: (1) non-septate hyphae, (2) wide hyphae with thick walls, and (3) absence of yeast phase. Non-septation of hyphae refers to a type of hyphal growth wherein walls (i.e. septations), do not separate individual hyphal cells, and in which multiple nuclei float in the filamentous hyphae (i.e. coenocytic growth).

Most of the pathogenic zygomycotes are non-commensal opportunists. They grow in soil, water, or air, on plants or on dung. Humans are constantly being exposed to their infective spores, by inhalation or by ingestion. Virtually all infections occur in patients who provide these fungi with a physiologic opportunity for growth (e.g., malnutrition, diabetes, advanced cancer, immunodeficiency, or an infection portal such as an indwelling catheter or a intravenous line).

Zygomycota
  Mucormycotina
    Mucorales
      Mucoraceae
        *Rhizopus
        *Mucor
        *Absidia
      Syncephalastraceae
        *Syncephalastrum
  Entomophthoramycotina
    Entomophthorales
      Basidiobolaceae
        *Basidiobolus
      Ancylistaceae
        *Conidiobolus

The pathogenic members of Class Zygomycota belong to one of two subclasses: Class Mucorales or Class Entomophthoramycotina. Infection with any zygomycote is known as zygomycosis. When the infectious agent is known to be a member of Class Mucorales, the disease is more specifically known as mucormycosis. Class Mucorales account for the bulk of infections caused by zygomycotes. Regardless of the Mucorales species, the clinical infections are similar. Common primary sites of infection are lungs, gastrointestinal tract, kidneys, and skin. Sinus infections, spreading to the nasopharynx, eyes, and brain, seem to have a particular affinity for diabetic individuals. Primary infections tend to be invasive, and may lead to disseminated disease. Rare infectious genera in Class Mucorales, aside from those listed here, have recently been isolated: Cokeromyces, Saksenaea, Apophysomyces, and Chlamydoabsidia [102].

Infections caused by genera of Class Entomophthorales produce a somewhat different clinical picture than that of Class Mucorales. The entomophthoramycoses are most often primary skin infections, and they can occur in immune-competent hosts. Infections caused by Genus Basidiobolus often arise on the trunk and thighs. Infections from members of Genus Conidiobolus typically arise on the nose and face [104].

Infectious species:

Mucorales:
- *Rhizopus oryzae* (mucormycosis, zygomycosis)
- *Mucor indicus* (mucormycosis, zygomycosis, phycomycosis)
- *Mucor racemosus* (mucormycosis, allergic skin reaction)
- *Absidia corymbifera* (mucormycosis, zygomycosis)
- *Syncephalastrum racemosum* (mucormycosis, zygomycosis, nail infection)

Entomophthorales:
- *Basidiobolus ranarum* (basidiobolomycosis, zygomycosis, entomophthoramycosis)
- *Conidiobolus coronatus* (conidiobolomycosis, zygomycosis, entomophthoramycosis)
- *Conidiobolus incongruus* (conidiobolomycosis, zygomycosis, entomophthoramycosis)

Chapter 35

Basidiomycota

"Only theory can tell us what to measure and how to interpret it."

Albert Einstein

Eukaryota
  Bikonta, 2-flagella (Chapter 16–21)
  Unikonta, 1-flagellum
    Amoebozoa (Chapter 22)
    Opisthokonta
      Choanozoa (Chapter 23)
      Animalia (Chapters 25–32)
      Fungi
        Zygomycota (Chapter 34)
          Entomophthoromycotina
          Mucoromycotina
        Dikarya
          Basidiomycota (Chapter 35)
            Agaricomycotina
              Tremellomycetes
                Tremellales
                  Tremellaceae
                    *Cryptococcus
            Ustilaginomycotina
              Exobasidiomycetes
                Malasseziales
                  Malasseziaceae
                    *Malassezia
          Ascomycota (Chapter 36)
        Microsporidia (Chapter 37)

Pathogenic subclasses of organisms belonging to Class Fungi can be divided into those subclasses that form dikaryons (i.e. Class Basidiomycota, Chapter 35, and Class Ascomycota, Chapter 36), and those subclasses that

J.J. Berman: Taxonomic Guide to Infectious Diseases. DOI: http://dx.doi.org/10.1016/B978-0-12-415895-5.00035-0

do not form dikaryons (Class Zygomycota, Chapter 34 and Class Microsporidia, Chapter 37). Dikaryons occur exclusively in the ascomycotes and basidiomycotes (i.e. in no other earthly organisms).

A dikaryon is a cell with a double nucleus, composed of two haploid nuclei that came to occupy the same cell, through conjugation, and without fusion of the two nuclei (i.e. two cells fused, but the nuclei within the cells do not fuse). A dikaryon is a dividing cell, wherein both nuclei divide synchronously, and both nuclei are metabolically active. The somatic cells of hyphae are haploid (from the Greek, "haplous," meaning single), having a complete set of unpaired chromosomes. Likewise, the gametes of fungi are haploid. Dikaryons can be formed by the fusion of haploid cells from two physically adjacent compatible mycelia (i.e. hyphae), or from the sexual fusion of two gametes, or from the fusion of a gamete with a haploid somatic cell. The dikaryotic state may be very short, or relatively long, but it eventually leads to a fused, diploid state. Diploid cells can yield, through meiosis, two haploid spore cells.

The two subclasses of Class Dikarya (Class Basidiomycota and Class Ascomycota) each have their own characteristic sexual and asexual bodies that produce cells that leave the organism and enter the environment, often as airborne spores. In Class Basidiomycota organisms reproduce sexually using a club-shaped structure, called a basidium, that produces basidiospores. They can also reproduce asexually by producing hardened spores from specialized cell structures (conidiophores), extending from hyphae.

A deep understanding of dikaryon biology and sexual reproduction is not particularly relevant to clinical mycology, as the tissue growth of virtually all infectious fungi is vegetative (i.e. yeast dividing by a budding process, or hyphae forming mycelium). Suffice it to say that the air we breathe, and, to some extent, the food we eat and the water we drink, carries the sexual and asexual spores of Class Basidiomycota.

Basidiomycota (Chapter 35)
  Agaricomycotina
    Tremellomycetes
      Tremellales
        Tremellaceae
          *Cryptococcus
  Ustilaginomycotina
    Exobasidiomycetes
      Malasseziales
        Malasseziaceae
          *Malassezia

Only two genera account for the human infections caused by members of Class Basidiomycota: Genus Cryptococcus and Genus Malassezia.

Cryptococcus species grow in tissues exclusively as yeasts (i.e. hyphal mycelia are never observed). *Cryptococcus neoformans* causes cryptococcal meningitis or meningo-encephalitis in immune-deficient patients. Occasional cases of cryptococcal meningitis are caused by *Cryptococcus laurentii* and *Cryptococcus albidus*.

Cryptococcus gattii produces highly virulent infections that involve the lungs, the meninges, or the brain. Until recently, it was confined to tropical or subtropical climates, and was particularly frequent in Papua New Guinea and northern Australia. In recent years, a few hundred cases have occurred in the northwest United States and in Vancouver. Unlike *Cryptococcus neoformans*, which causes disease almost exclusively in immune compromised individuals, *Cryptococcus gattii* occurs in immune competent individuals.

Malassezia species produce a type of ringworm called tinea versicolor (alternately known as pityriasis versicolor), and a folliculitis called pityrosporum folliculitis (Malassezia was formerly known as Pityrosporum). Mallasezia species are part of normal skin flora. Active infections, when they occur, arise from endogenous skin organisms, not through contagion with infected humans. Tinea versicolor occurs in up to 8% of the general population, most frequently in adolescents. It consists of a round itchy macular rash with a sharply demarcated circumference; hence, falling into the broad clinical category of so-called ringworm fungal infections. In biopsies of infected skin, yeast forms admixed with hyphae are seen, producing a histologic appearance likened to spaghetti and meatballs. Most cases of tinea versicolor are caused by *Malassezia globosa.* In pityrosporum folliculitis, organisms descend into hair follicles, producing inflammation, with papule formation.

The clinical term "ringworm" is irreconcilable with taxonomic nomenclature, and a few words of explanation are required. Ringworm is synonymous with dermatophytosis, skin infections caused by one of several fungi that live in the top, keratin layer of the epidermis, producing round macules or plaques. Within Class Ascomycota (Chapter 36), there is a subclass called Arthrodermataceae that contains the traditional dermatophytic genera: Epidermophyton, Microsporum, and Trichophyton. Nonetheless, additional fungal genera, outside Class Arthrodermataceae may produce ringworm infections: Hortaea (an ascomycote), and Malassezia (a basidiomycote).

Malassezia species are ubiquitous and grow, in yeast form, as commensals on normal keratinized skin. They require fatty acids, and are thus found in highest concentrations in sebum-rich areas, including the face. They have been found in a high percentage of cases of several common and mild skin disorders, including dandruff, seborrheic dermatitis, and even hyperhidrosis. The pathogenic role of Malassezia species in these diseases is obscure.

As with most fungal infections, otherwise mild conditions can progress into life-threatening diseases in individuals who are malnourished or immune-deficient. Malassezia species have been involved in serious fungal infections arising in low-birth-weight infants. Additional cases have occurred in adults who receive intravenous parenteral nutrition, presumably through the introduction of fungi via intracatheter growth on lipid-rich alimentation fluids [105].

Infectious species:

Cryptococcus neoformans (cryptococcal meningitis)
Cryptococcus gattii (pulmonary cryptococcosis, basal meningitis, and cerebral cryptococcomas)
Malassezia globosa (tinea versicolor, pityrosporum folliculitis)
Malassezia ovale, fomerly *Pityrosporum ovale* (tinea versicolor, pityrosporum folliculitis)
Malassezia furfur (fungemia in low-birth-weight neonates)
Malassezia pachydermatis (fungemia in low-birth-weight neonates)

Chapter 36

Ascomycota

"One does not discover new lands without consenting to lose sight of the shore for a very long time."

Andre Gide

Eukaryota
  Bikonta, 2-flagella (Chapter 16–21)
  Unikonta, 1-flagellum
    Amoebozoa (Chapter 22)
    Opisthokonta
      Choanozoa (Chapter 23)
      Animalia (Chapters 25–32)
      Fungi
        Zygomycota (Chapter 34)
          Entomophthoromycotina
          Mucoromycotina
        Dikarya
          Basidiomycota (Chapter 35)
          Ascomycota (Chapter 36)
            Saccharomycotina
              Saccharomycetes
                Saccharomycetales
                  Saccharomycetaceae
                    *Candida
            Taphrinomycotina
              Pneumocystidomycetes
                Pneumocystidales
                  Pneumocystidaceae
                    *Pneumocystis
            Pezizomycotina
              Dothideomycetes
                Dothideales
                  Dothioraceae

J.J. Berman: Taxonomic Guide to Infectious Diseases. DOI: http://dx.doi.org/10.1016/B978-0-12-415895-5.00036-2

            *Hortaea
        Pleosporomycetidae
            Pleosporales
                Testudinaceae
                    *Neotestudina
    Sordariomycetes
        Microascales
            Microascaceae
                *Scedosporium
        Hypocreales
            Nectriaceae
                *Fusarium
        Sordariales
            Incertae sedis (uncertain)
                *Madurella
    Eurotiomycetes
        Eurotiales
            Trichocomaceae
                *Aspergillus
                *Penicillium
        Herpotrichiellaceae
            *Fonsecaea
            *Cladophialophora
            *Phialophora
        Onygenales
            Ajellomycetaceae
                *Emmonsia
                *Histoplasma
                *Blastomyces [106]
                *Paracoccidioides
            Arthrodermataceae (the dermatophytes)
                *Epidermophyton
                *Microsporum
                *Trichophyton
            Onygenaceae
                *Coccidioides
            Incertae sedis (unknown)
                *Lacazia
    Ophiostomatales
        Ophiostomataceae
            *Sporothrix
Microsporidia (Chapter 37)

There are four classes of fungi that contain pathogenic organisms: Class Zygomycota (Chapter 34), Class Basidiomycota (Chapter 35), Class Ascomycota (Chapter 36), and Class Microsporidia (Chapter 37). Class Ascomycota contains the greatest number of fungal organisms infectious in humans, and it contains most of the fungi that regularly cause clinically life-threatening disease in otherwise healthy individuals. Class Ascomycota, along with Class Basidiomycota, comprise the dikaryotic fungi (see Chapter 35 for full discussion).

All members of Class Ascomycota that reproduce sexually produce an ascus (from the Greek "askos," meaning sac), containing spores. Unfortunately for taxonomists, many members of Class Ascomycota simply do not reproduce sexually; hence, they do not produce the ascus that characterizes their taxonomic class. Taxonomists invented a temporary class of organisms known as the deuteromycotes (or imperfect fungi) to hold these asexual species. Thanks to molecular analyses, many of these ascus-impaired species have been sorted into proper subclasses within Class Ascomycota. Currently, three major classes account for all of the pathogenic members of Class Ascomycota: Saccharomycotina, Taphrinomycotina, and Pezizomycotina. Class Saccharomycotina are yeasts; round, unicellular fungi that reproduce by budding. This class contains a single genus that is pathogenic in humans: Candida. Class Taphrinomycotina contains a single species that is pathogenic in humans: *Pneumocystis jiroveci.* All of the remaining Ascomycotes, and there are many, belong to Class Pezizomycotina.

Ascomycota
  Saccharomycotina
    Saccharomycetes
      Saccharomycetales
        Saccharomycetaceae
          *Candida

Candida, the sole pathogenic genus in Class Saccharomycotina, is a normal inhabitant of humans, and various species are found on the skin, respiratory tract, gut, and female genital tract of healthy individuals. An ecological balance exists between Candida species and various bacterial commensals. When this balance is disrupted by the use of antibiotics, overgrowth of Candida species may occur. In addition, as with virtually all of the pathogenic fungi, overt diseases may occur in immune-deficient individuals. Patients undergoing intense chemotherapy are at particular risk for life-threatening candidal infections.

The least worrisome of the candidiases are superficial infections confined to mucosal surfaces. These are common in the mouth and GI tract and are characterized by thick colonies of yeast that form a white surface crust. So-called invasive candidiasis involves the penetration of organisms through the mucosa into deeper tissue layers, and the transition to invasiveness is often accompanied by a change in morphology from the yeast form, to elongated cells (pseudohyphae) and hyphae. The most serious stage of candidiasis involves growth in blood (candidemia) and dissemination to distant organs. *Candida albicans* is the most common pathogenic species, but there are many more known pathogenic types, including *Candida dubliniensis*, *Candida glabrata*, *Candia parapsilosis*, *Candida rugosa*, and *Candida tropicalis*.

Ascomycota
  Taphrinomycotina
    Pneumocystidomycetes
      Pneumocystidales
        Pneumocystidaceae
          *Pneumocystis

Pneumocystis, the sole pathogenic genus in Class Taphrinomycotina, was, until recently, presumed to be a protozoa. Early papers invented a detailed protozoan life cycle for Pneumocystis, complete with morphologically distinct developmental stages, that included cyst, trophozoite, sporozoite, and intracystic bodies [107]. To be fair to the early taxonomists, the classification of Pneumocystis was particularly difficult because the organism could not be grown in culture. Owing to molecular analyses, we now know that Pneumocystis is a fungus that grows as a yeast. The so-called trophozoite stage of Pneumocystis is equivalent to the vegetative stage of well-studied *Schizosaccharomyces pombe*, a non-pathogenic member of Class Taphrinomycotina. The yeasts form an enclosed cyst, which eventually ruptures, releasing spores. These different forms of Pneumocystis comprise the various morphologic forms of the fungus that are seen in histologic sections of infected lungs. *Pneumocystis jerovicii* (formerly *Pneumocystis carinii*), produces pneumonia in immune-deficient individuals. AIDS patients are particularly vulnerable to Pneumocystis infections.

Ascomycota
  Pezizomycotina
    Dothideomycetes
      Dothideales
        Dothioraceae
          *Hortaea
      Pleosporomycetidae
        Pleosporales
          Testudinaceae
            *Neotestudina

Hortaea werneckii causes tinea nigra, from the Latin "tinea," meaning worm and "niger," meaning black. Tinea, also known as ringworm, is a localized infection of the keratin layer of the skin. The tinea lesions produced by *Hortaea wernickii* are black because the organism produces melanin. Melanin production is a feature that can help clinical mycologists identify a fungal species. Species that produce melanin are called dematiaceous fungi.

Neotestudina is one of various genera that has been cultured from mycetoma, an uncommon and enigmatic skin infection. Mycetoma, also known as Madura foot and maduromycosis, occurs most often in India, Africa, and South America. It presents as a slowly growing, fungating mass arising in the subcutaneous tissues, usually of the foot. As the mass grows, draining sinuses discharge fluid and hard grains (white, white-yellow, or black grains). These masses often become superinfected, making it very difficult to determine the primary pathogen that causes the disease. More than thirty different species of bacteria and fungi have been grown from these lesions. It has been claimed that black grain mycetomas is caused by *Leptosphaeria senegalensis*, *Madurella grisea*, *Madurella mycetomatis*, or *Pyrenochaeta romeroi*. White grain mycetomas are reputedly caused by Acremonium species, *Aspergillus nidulans*, *Neotestudina rosatii*, or *Pseudallescheria boydii*. White-yellow grain mycetomas are said to be caused by: *Actinomadura madurae*, *Nocardia asteroides*, and *Nocardia brasiliensis*. Brown-red grain mycetomas are said to be caused by: *Actinomadura pelletieri* or *Streptomcyes somaliensis*. Taken at face value, these claims would indicate that many different organisms, both bacterial and fungal, can produce a disease of remarkably specific, even unique, clinical features. Suffice it to say that clinical science has much to learn about mycetoma.

Ascomycota
  Pezizomycotina
    Sordariomycetes
      Microascales
        Microascaceae
          *Scedosporium
      Hypocreales
        Nectriaceae
          *Fusarium
      Sordariales
        Incertae sedis (uncertain)
          *Madurella

Scedosporium prolificans accounts for a sizable portion of the instances of an uncommon lesion: disseminated phaeohyphomycosis. The prefix derives from the Greek "phaeo," meaning dusky. The suffix "hyphomycosis" indicates that the fungal organism produces hyphae. Phaeohyphomycosis presents as one or more abscesses that are brown, on gross examination. The fungi that cause phaeohyphomycosis (e.g. *Scedosporium prolificans*) are dematiaceous (melanin-producing). Other genera of dematiceious fungi that may cause phaeohyphomycotic abscesses in immune-compromised patients include Alternaria, Curvularia, Phialophora, and Cladiophora (see below).

Species of Fusarium can cause corneal keratitis and onychomycosis (fungal nail infection) in otherwise healthy individuals. In immune-compromised patients with very low white blood cell counts, various Fusarium species can produce life-threatening disseminated infections. These species include *Fusarium oxysporum*, *Fusarium proliferatum*, *Fusarium solani*, and *Fusarium verticillioides*. Aside from their role in human infection, Genus Fusarium has been studied for its ability to produce various powerful mycotoxins. Poisoning outbreaks from Fusarium-contaminated food have been reported [108].

Madurella species were included in the discussion of Genus Neotestudina (see above). They are among the many putative causes of maduromycosis.

Ascomycota
  Pezizomycotina
    Eurotiomycetes
      Eurotiales
        Trichocomaceae
          *Aspergillus
          *Penicillium
      Herpotrichiellaceae
        *Fonsecaea
        *Cladophialophora
        *Phialophora
      Onygenales
        Ajellomycetaceae
          *Emmonsia
          *Histoplasma
          *Blastomyces [106]
          *Paracoccidioides
        Arthrodermataceae (the dermatophytes)
          *Epidermophyton
          *Microsporum
          *Trichophyton
        Onygenaceae
          *Coccidioides
        Incertae sedis (unknown)
          *Lacazia

Genus Aspergillus contains the ubiquitous species that cause aspergillosis. The most common species associated with aspergillosis is *Aspergillus fumigatus*. Aspergillosis was discussed in Chapter 33 (Overview of Class Fungi), in which the many different forms of fungal infection might manifest. In the case of aspergillosis, infections begin in the lung, and may produce colonization of the airways, without disease. Alternately, they may provoke an acute or chronic allergic reaction in the lungs. The organism can grow in respiratory mucosa, or it may invade into the lung tissue. It may produce large fungal masses, or it may invade diffusely through the lung, like a pneumonia. Or it may produce a fungemia, and disseminate throughout the body. Spores of Aspergillus species are found in the air, and everyone is exposed to these fungi. Disease most often occurs in immune-compromised individuals. Primary cutaneous aspergillosis is a rare form of aspergillosis that occurs in the skin of immune-compromised patients near the site of indwelling intravenous lines.

Genus Penicillium contains one species that is known to produce human disease: *Penicillium marneffei.* Infections occur primarily in Southeast Asia, and most reported cases have occurred in AIDS patients. The disease, known as penicillosis, can produce systemic disease. This species, like many of the other highly pathogenic members of Class Ascomycota, is a dimorphic fungus.

Chromoblastomycosis can be caused by a variety of organisms that belong to Class Herpotrichiellaceae: *Fonsecaea pedrosoi*, *Phialophora verrucosa*, *Cladophialophora carrionii*, and *Fonsecaea compacta.* Chromoblastomycosis begins as a skin papule, at the site of entry, and over the years may slowly spread.

Cladophialophora bantiana can produce phaeohyphomycotic brain abscesses and subcutaneous lesions in both normal and immunosuppressed patients.

Class Onygenales contains many of the fungi that characteristically cause disease in immune-competent individuals, and it is the only class of fungi containing organisms that infect, disseminate, and sometimes kill otherwise healthy persons, with one important qualification. Good health is hard to establish with certainty, and the biological relationships between human host and fungal infection can be very complex. Although most pathogenic fungi produce clinical disease in immune-deficient individuals, you will occasionally encounter a supposedly normal person who develops a fulminant infection from a supposedly non-pathogenic or opportunistic fungus. Nonetheless, there are a few fungi that produce disease in healthy patients, and most of these belong to Class Onygenales. *Paracoccidioides brasiliensis* (paracoccidioidomycosis or South American blastomycosis), along with *Coccidioides immitis* (coccidioidomycosis or valley fever), *Blastomyces dermatitidis* (blastomycosis), and *Histoplasma capsulatum* (histoplasmosis), are highly pathogenic. Each is characterized by the growth of yeast (round cells) in diseased tissues, and each is dimorphic in culture medium. All four are systemic mycoses that begin as lung infections. All four grow in the soil or, in the case of *Histoplasma capsulatum*, in bat or bird guano. Each has its own geographic distribution. Paracoccidioidomycosis is found in South America; coccidioidomycosis in the Southwestern USA, blastomycosis in the Midwest and Northern USA, and histoplasmosis in the central and eastern North America.

Lacazia loboi, formerly *Loboa loboi*, is another species in Class Onygenales that causes disease in otherwise healthy individuals. The disease, endemic to the Amazon, is known by a number of names: Lobo disease, lacaziosis, keloidal blastomycosis, Amazonian blastomycosis, miraip, piraip, and lobomycosis. Clinically, Lobo's disease is a granulomatous infection of the skin. The disease can be mistaken for *Paracoccidioides brasiliensis* and with *Blastomyces dermatididis* due to the similar morphology of the yeast in tissues. *Lacazia loboi* has not been successfully cultured. Like most fungal infections, transmission comes from the environment, not through human contagion.

Genus Emmonsia contains two pathogenic organism: *Emmonsia parva*, alternately known as *Chrysosporium parvum* var. *parvum*, and *Emmonsia crescens*, alternately known as *Chrysosporium parvum* var. *crescens*. These organisms are the causative agents of adiaspiromycosis, a disease with a unique pathogenic mechanism. Adiaspiromycosis causes pulmonary disease in various animal species, particularly rodents. It is a rare cause of disease in humans. Although it is referred to as an infection, it is actually a foreign body reaction, resulting from the inhalation and sequestration of conidial spores in the small branches of the respiratory tree. The spores are large, about 300 microns in diameter. Histologic cross-sections of infected lungs show a walled spore, surrounded by acute and chronic inflammation and foreign body reaction granulomas. There is no growth of the organism. In the literature, the term "disseminated adiaspiromycosis" is sometimes encountered [109], referring to lesions that involve most of the lung parenchyma. In this case, dissemination is not an indication the lesion has spread from one part of the lung to another, but that the load of inhaled spores involves most of the respiratory tree.

Class Onygenales also contains Class Arthrodermataceae, the dermatophytes: Genus Epidermophyton, Genus Microsporum, and Genus Trichophyton. Species within these genera account for most of the tinea, also known as ringworm, infections of humans. Exceptions are *Hortaea werneckii* (an ascomycote in Class Dothideomycetes), the cause of tinea nigra; and *Malassezia furfur* (a basidiomycota), the cause of tinea versicolor. All tinea infections are colonizations of the superficial layers of the epidermis by keratin-loving fungi. Like other infections caused by members of Class Onygenales, tinea infections can occur in immune-competent hosts. Like most fungal infections, the clinical features of the disease tend to worsen in immune-compromised patients. In immune-compromised patients, a case of superficial tinea may progress into a locally invasive process (tinea profunda) or recurrent infections [110].

As previously described, fungal diseases are not typically spread from person to person. They are spread by fungi that grow in the environment and release infective spores into air, soil, or water. In the case of the dermatophytes, disease is typically spread when infected persons shed fungi into shared areas that support the growth of the fungi (e.g. wet bathroom floors, moldy towels, shared sandals). Direct transmission from person to person or animal to person (e.g. tinea due to *Microsporum canis* infections on cats and dogs) is also possible.

It is important not to confuse Microsporum with the fungus of Genus Microsporidium (Class Microsporidia Chapter 37).

Ascomycota
  Pezizomycotina
    Ophiostomatales
      Ophiostomataceae
        *Sporothrix

Sporotrichosis is caused by *Sporothrix schenckii*, a fungus that grows in soil, particularly peat moss. Gardeners who handle soil and peat are infected through abrasions on their hands. The lesion begins as a localized mass. As time passes, the satellite lesions advance up the arm, via lymphatic spread. Sporotrichosis is endemic to Peru.

Infectious species:

Emmonsia parva, also known as *Chrysosporium parva* (adiaspiromycosis or haplomycosis)
Emmonsia crescens, also known as *Chrysosporium crescens* (adiaspiromycosis or haplomycosis)
Fusarium species (corneal keratitis)
Histoplasma capsulatum (histoplasmosis)
Histoplasma duboisii (African histoplasmosis)
Fonsecaea compacta (chromoblastomycosis, also known as chromomycosis, and cladosporiosis)
Neotestudina rosatii (white grain mycetoma)
Cladophialophora bantiana, formerly *Xylohypha bantiana*, formerly *Cladosporium bantianum*) (cerebral phaeohyphomycosis)
Aspergillus fumigatus (aspergillosis)
Penicillium marneffei (penicilliosis)
Epidermophyton floccosum (athlete's foot, tinea pedis, tinea cruris, tinea corporis, onychomycosis)
Microsporum canis (ringworm, dermatophytosis, tinea)
Trichophyton rubrum (ringworm, dermatophytosis, tinea)
Trichophyton tonsurans (ringworm, dermatophytosis, tinea, 90% of cases of tinea capitis in North America, tinea favosa)
Blastomyces dermatitidis, *Ajellomyces dermatitidis* (blastomycosis), see *Lacazia loboi*, with which it may be mistaken
Pneumocystis jiroveci, formerly *Pneumocystis carinii* (Pneumocystis pneumonia)

Coccidioides immitis (coccidioidomycosis, also known as San Joaquin valley fever or valley fever)
Paracoccidioides brasiliensis (paracoccidioidomycosis, or South American Blastomycosis or Brazilian Blastomycosis)
Scedosporium apiospermum, *Pseudallescheria boydii* (lung disease, disseminated infection, mycetoma, in immune-compromised individuals)
Scedosporium proliferans (inflatum)
Sporothrix schenckii (sporotrichosis, rose-handler's disease)
Candida albicans (candidiasis, candidemia, thrush)
Candida glabrata (urinary tract infection and sepsis in immune-compromised individuals)
Candida parapsilosis (wound infection and sepsis in immune-compromised individuals)
Candida dubliniensis (infections in immune-compromised individuals)
Candida tropicalis (frequent cause of sepsis and disseminated candidiasis in immune-compromised individuals)
Lacazia loboi (Jorge Lobo disease, Lobo disease, lacaziosis, keloidal blastomycosis, Amazonian blastomycosis, blastomycoid granuloma, miraip, piraip, lobomycosis)
Hortaea werneckii (tinea nigra)

Chapter 37

Microsporidia

"The most savage controversies are those about matters as to which there is no good evidence either way."

Bertrand Russell

Eukaryota
  Bikonta, 2-flagella (Chapters 16–21)
  Unikonta, 1-flagellum
    Amoebozoa (Chapter 22)
    Opisthokonta
      Choanozoa (Chapter 23)
      Animalia (Chapters 25–32)
      Fungi
        Zygomycota (Chapter 34)
        Dikarya
          Basidiomycota (Chapter 35)
          Ascomycota (Chapter 36)
        Microsporidia (Chapter 37)
          Encephalitozoonidea
            *Encephalitozoon
          Enterocytogoonidea
            *Enterocytozoon
          Microsporidea
            *Microsporidium
          Nosematidea
            *Brachiola
            *Nosema
            *Vittaforma
          Pleistophoridea
            *Pleistophora
            *Trachipleistophora

J.J. Berman: Taxonomic Guide to Infectious Diseases. DOI: http://dx.doi.org/10.1016/B978-0-12-415895-5.00037-4

Class Microsporidia is not your typical fungus. Unlike all other fungal classes, the members of Class Microsporidium are obligate intracellular parasites, that have adapted themselves to parasitic lives in a wide range of eukaryotic organisms. Unlike virtually all other members of Class Fungi, the members of Class Microsporidia lack mitochondria. As with other so-called amitochondriate eukaryotic classes (e.g. Class Metamonada, Chapter 16 [65], and Class Amoebozoa, Chapter 22 [66]), the members of Class Microsoridia have retained remnant forms of mitochondria (i.e. mitosomes, hydrogenosomes) [67]. Unlike most other fungi, the microsporidia lack a hyphal form and do not produce multicellular tissue structures. With all these non-fungal properties, taxonomists never entertained the notion that the microsporidia were fungi; until recently. Several molecular phylogenetic studies suggest that microsporidians are fungi [111].

All members of Class Microsporidia form spores, thick-walled cells that can survive in the environment. Infected animals pass spores in their urine and feces, and the spores infect humans by contact or inhalation. The spores pass to the intestines, where they extrude a polar tube into the intestinal lining cells. Through the polar tube, the cytoplasm of the spore (sporoplasm) enters the host cell and organizes into a cell capable of division. Eventually, more spores are formed. When the host cells lyse (break open and die), spores are released into the intestine, and are passed with feces into the environment.

A wide variety of animals are reservoirs for the various species of Microsporidia: mammals, birds, and insects. The spores are passed in the stools, and infect humans through direct contact, water contamination, or through respiration of airborne spores. Preliminary evidence suggests that microsporidial infections are common [112]. Most, but not all, cases of symptomatic microsporidiosis occur in immune-compromised individuals, particularly in patients who have AIDS.

Although microsporidiosis is considered a rapidly emerging disease, we lack important and fundamental epidemiologic information. How prevalent is the organism in the immune-competent population? How prevalent is the organism in the population of immune-deficient but asymptomatic patients? How often is a microsporidium the causative agent of diarrheal diseases among different age groups? In which geographic regions does microsporidiosis occur? What are the most important animal reservoirs for human microsporidiosis?

Microsporidia
    Encephalitozoonidea
        *Encephalitozoon
    Enterocytogoonidea
        *Enterocytozoon

Microsporidea
*Microsporidium
Nosematidea
*Brachiola
*Nosema
*Vittaforma
Pleistophoridea
*Pleistophora
*Trachipleistophora

At least 14 species of Microsporidia, in eight genera, can cause microsporidiosis. Most produce symptoms of diarrhea that can be chronic, and wasting, in susceptible individuals. Some microsporidian species produce a variety of additional conditions, associated with ocular infections, muscle infections, and even systemic disease.

Readers should remember not to confuse Microsporidia with Microsporum, a fungal genus causing dermatophytosis. It is also important not to confuse microsporidiosis with cryptosporidiosis, an apicomplexan disease (Apicomplexa, Chapter 19), that also produces diarrhea in immune-compromised patients.

Infectious species:

Brachiola algerae (microsporidiosis, keratoconjunctivitis, skin and deep muscle infection)
Brachiola connori (microsporidiosis, ocular infection)
Brachiola vesicularum (microsporidiosis)
Encephalitozoon cuniculi (microsporidiosis, keratoconjunctivitis, infection of respiratory and genitourinary tract, disseminated infection)
Encephalitozoon hellem (microsporidiosis, keratoconjunctivitis, infection of respiratory and genitourinary tract, disseminated infection)
Encephalitozoon intestinalis, formerly *Septata intestinalis* (microsporidiosis, infection of the GI tract causing diarrhea, and dissemination to ocular, genitourinary and respiratory tracts)
Enterocytozoon bieneusi (microsporidiosis, diarrhea, acalculous cholecystitis)
Microsporidium ceylonensis (microsporidiosis, infection of the cornea)
Microsporidium africanum (microsporidiosis, infection of the cornea)
Nosema ocularum (microsporidiosis, ocular infection)
Pleistophora sp. (microsporidiosis, muscular infection)
Trachipleistophora hominis (microsporidiosis, muscular infection, keratitis)
Trachipleistophora anthropophthera (microsporidiosis, disseminated infection)
Vittaforma corneae, same as *Nosema corneum* (microsporidiosis, urinary tract infection, ocular infection)

Part VI

Nonliving Infectious Agents: Viruses and Prions

Chapter 38

Overview of Viruses

"The human genome is a living document of ancient and now extinct viruses."

Michael Emerman and Harmit Malik [113]

In this book, viruses and prions are referred to as "biological agents"; not as living organisms. Viruses lack key features that distinguish life from non-life. They depend entirely on host cells for replication; they do not partake in metabolism, and do not yield energy; they cannot adjust to changes in their environment (i.e. no homeostasis), nor can they respond to stimuli. Most scientists consider viruses to be mobile genetic elements that can travel between cells (much as transposons are considered mobile genetic elements that travel within a cell). All viruses have a mechanism that permits them to infect cells and to use the host cell machinery to replicate. At a minimum, viruses consist of a small RNA or DNA genome, encased by a protective protein coat, called a capsid. Some viruses carry one or several proteins. Some viruses have an envelope extracted, in part, from host cell membranes.

For non-living organisms dependent entirely on host cells for their continued existence, viruses have done extremely well for themselves. Every class of living organism hosts viruses. Viruses are literally everywhere in our environment, and are the must abundant life form in the oceans, in terms of numbers of organisms [114]. At least 8% of the human genome is composed of fragments derived from RNA viruses, acquired in our genetic past [115]. As far as anyone knows, viruses are as ancient as the earliest forms of terrestrial life.

When did the first viruses appear on earth? Nobody knows, but much of the current speculation on the origins of viral life centers on the so-called RNA world, wherein the biogenic precursors of living organisms held domain in ancient oceans. Before there were membrane-enclosed organisms, there were volcanic rocks. When basalt melts and aggregates, small bubbles, about the size of bacteria, form and interconnect with one another [116]. Examples of porous volcanic rocks include tufa, tuff, travertine, pumice, stromboli basalt, and scoria. Any of these water-drenched rocks could have been crucibles for the earliest biogenic molecules. RNA almost certainly preceded the appearance of DNA, as DNA bases require synthesis from RNA

J.J. Berman: Taxonomic Guide to Infectious Diseases. DOI: http://dx.doi.org/10.1016/B978-0-12-415895-5.00038-6

nucleotides via the action of ribonucleotide reductase and thymidylate synthetase. RNA molecules may have developed their biogenic properties, as templates for synthesizing proteins, within rock bubbles. Silicates and hard surfaces in rock may have served as structural catalysts, holding molecules in place as they were lengthened (RNA synthesis), replicated (RNA replication), and translated (protein synthesis) from precursor molecules found in sea-water. Because rock bubbles are interconnected, early RNA virus-like molecules could move from bubble to bubble, exchanging genetic materials.

Sometime later, DNA may have appeared. DNA is a much more stable molecule than RNA, less prone to replication error, and less susceptible to intrusion by RNA viruses that were freely commuting between rock bubbles. The evolution of DNA modifications (adherent proteins and base methylations, characterizing the early epigenome) may have developed as a defense against infection by RNA viruses.

What came next? After DNA appeared, it was inevitable that DNA viruses would emerge. DNA viruses, being more stable than RNA viruses, could grow into large, complex entities, such as the megaviruses (Group I double-stranded DNA Viruses, Chapter 39). Then came cell membranes. It is known that phospholipids spontaneously form lipid bilayers in agitated water. It seems plausible that rock-dwelling organisms, endowed with an enclosing bilayer membrane assembled from phosphorylated small molecules would eventually venture out into the ocean, searching for the best sources of food. The late emergence of enclosing membranes, well after the initial development of membrane-less organisms, is supported by the profound structural differences in bacterial and archaean membranes [117]. If the two classes of organisms had split off from a common, membrane-enclosed ancestor, you might expect them to have similar membranes.

Hypothesizing on the origins of living and non-living organisms is all good sport, but it should not be taken too seriously. Suffice it to say that theorists have proposed that all life on earth developed from ancient viruses [118,119]. Motivated readers can delve into the subject and draw their own conclusions.

At this time, the classification of viruses is somewhat crude. Anything you choose as a classifying principle fails to biologically unify the subclasses. For example, if you classify viruses by their genomic molecules (i.e. DNA or RNA, single strandedness or double strandedness), you will find that subclasses of the same genomic type, will have dissimilar structures: envelope, size, shape, proteins, capsid. When we list viruses based on method of contagion, by persistence within host (i.e. acute, chronic, latent, or persistent), toxicity (lytic, immunogenic), or by target cell specificity, no consistent taxonomic correlation is found.

Though we cannot classify viruses phylogenetically, at this time, we can usefully group viruses based on the physical characteristics of their genomes. The Baltimore Classification divides viruses into seven groups based on whether their genome is DNA, RNA, single-stranded or double-stranded, the sense of the single strand, and the presence or absence of a reverse transcriptase. The following schema shows the classes and subclasses of pathogenic viruses that will be described in Chapters 39 through 45.

Group I, double-stranded DNA (Chapter 39)
Group II, single-stranded DNA (Chapter 40)
Group III, double-stranded RNA (Chapter 41)
Group IV, positive sense single-stranded RNA (Chapter 42)
Group V, negative sense single-stranded RNA (Chapter 43)
Group VI, single-stranded RNA with a reverse transcriptase (Chapter 44)
Group VII, double-stranded DNA with a reverse transcriptase (Chapter 45)

Chapter 46 will describe the infectious protein agents known as prions. Prions may be considered deformed proteins that have stumbled upon a way to infect cells and cause other proteins, of the same kind, to deform (i.e. non-synthetic replication).

It is worth repeating that when we use the Baltimore Classification (or any alternate viral classification, for that matter) we must grudgingly accept the fact that biologically relevant features of grouped viruses will cross taxonomic boundaries. Consider the arboviruses. Arbovirus is a shortened name for Arthropod borne virus. The arboviruses fall into several different groups of viruses. The principal vectors of the arboviruses are mosquitoes and ticks. Mosquito-borne arboviruses are members of Class Bunyaviridae (Group V), Flaviviridae (Group IV), or Togaviridae (Group IV). Tick-borne arboviruses are members of Class Bunyarviridae (Group V), Flaviviridae (Group IV), or Reoviridae (Group III). Over 500 arboviruses, infecting a variety of animals, have been described [76]. The lists of arboviruses, organized by vector, shown below, follow no taxonomic principle.

Mosquito-borne viruses
- Bunyaviridae (Group V, Chapter 43)
  - La Crosse encephalitis virus
  - California encephalitis virus
  - Rift Valley fever virus
- Flaviviridae (Group IV, Chapter 42)
  - Japanese encephalitis virus
  - Australian encephalitis virus
  - St. Louis encephalitis virus
  - West Nile fever virus
  - Dengue fever virus
  - Yellow fever virus
  - Zika fever virus
- Togaviridae (Group IV, Chapter 42)
  - Eastern equine encephalomyelitis virus

Western equine encephalomyelitis virus
Venezuelan equine encephalomyelitis virus
Chikungunya virus
O'Nyong-nyong fever virus
Ross River fever virus
Barmah Forest virus

Tick-borne viruses
Bunyaviridae (Group V, Chapter 43)
Crimean-Congo hemorrhagic fever virus
Flaviviridae (Group IV, Chapter 42)
Tick-borne encephalitis virus
Powassan encephalitis virus
Deer tick encephalitis virus
Omsk hemorrhagic fever virus
Kyasanur forest disease virus (Alkhurma virus)
Langat virus
Reoviridae (Group III, Chapter 41)
Colorado tick fever virus

The term "arbovirus" somewhat arbitrarily excludes non-arthropod vectors, such as rodents and bats.

Rodent-borne viruses (roboviruses)
Arenaviridae (Group V, Chapter 43)
Lassa fever
Venezuelan hemorrhagic fever (Guanarito virus)
Argentine hemorrhagic fever (Junin virus)
Bolivian hemorrhagic fever (Machupo virus)
Lujo virus
Bunyaviridae (Group V, Chapter 43)
Puumala virus
Andes virus
Sin Nombre virus
Hantavirus

Bat-borne viruses [120] (see Glossary item, Bat)
Filoviridae (Group V, Chapter 43)
Ebola hemorrhagic fever
Marburg hemorrhagic fever
Rhabdoviridae (Group V, Chapter 43)
Australian bat lyssavirus
Rabies virus
Mokola virus
Duvenhage virus

Lagos bat virus
Duvenhage virus

It would seem that we do not know enough about the origin and phylogeny of the different classes of viruses to create a true classification, wherein viruses of a class share a common set of inherited relationships. There is, however, hope for a better future. Highly innovative work in the field of viral phylogeny is proceeding along a variety of different approaches, including: inferring retroviral phylogeny by sequence divergences of nucleic acids and proteins in related viral species [121]; tracing the acquisition of genes in DNA viruses [122]; and dating viruses by the appearance of viral-specific antibodies in ancient host cells [113]. Because viruses evolve very rapidly, it is possible to trace the evolution of some viruses, with precision, over intervals as short as centuries or even decades [123]. It should be noted that before the advent of ribosomal sequence analysis, and as recently as the early 1970s, bacterial phylogeny was considered a hopeless field [21]. Bacteria were grouped by morphology, nutritional requirements, and enzymatic reactions (e.g. hemolysis, coagulase) without much attention to phylogenetic relationships. The field of viral phylogeny is quickly catching up with the phylogeny of living organisms.

Chapter 39

Group I Viruses: Double-Stranded DNA

"What trap is this? Where were its teeth concealed?"
Philip Larkin, from his poem "Myxomatosis"

Group I, dsDNA
  Herpesvirales
    Herpesviridae
      *Epstein–Barr virus
      *Herpes simplex virus type 1
      *Herpes simplex virus type 2
      *Herpes virus varicella-zoster virus
      *Herpesvirus simiae, also known as B virus
      *Human herpesvirus type 6, HHV6
      *Human herpesvirus type 7, HHV7
      *Human herpesvirus type 8, HHV8
      *Cytomegalovirus
  Unassigned
    Nonenveloped
      Adenoviridae
        *Human adeonviruses A through G
      Papillomaviridae
        *Human papillomavirus
      Polyomaviridae
        *BK polyomavirus
        *JC polyomavirus
        *Simian virus 40
    Nucleocytoplasmic large DNA viruses (NCLDV viruses)
      Poxviridae
        Orthopoxvirus
          *Buffalopox virus
          *Cowpox virus

J.J. Berman: Taxonomic Guide to Infectious Diseases. DOI: http://dx.doi.org/10.1016/B978-0-12-415895-5.00039-8

        *Monkeypox virus
        *Vaccinia virus
        *Variola major virus
        *Variola minor virus
      Parapoxvirus
        *Orf
        *Milker's node virus
      Molluscipoxvirus
        *Molluscum contagiosum virus
      Yatapoxvirus
        *Tanapoxvirus
        *Yaba monkey tumor virus
    Mimiviridae
      Mimivirus
        *Acanthamoeba polyphaga mimivirus (pneumonia)
Group II, ssDNA (Chapter 40)
Group III, dsRNA (Chapter 41)
Group IV (+)ssRNA (Chapter 42)
Group V (−)ssRNA (Chapter 43)
Group VI, ssRNA-RT (Chapter 44)
Group VII, dsDNA-RT (Chapter 45)
Prions (Chapter 46)

The Group I viruses all have a double-stranded DNA genome. Aside from this property, the viruses vary greatly. Some species have envelopes; others do not. Some species have circular genomes; others have linear genomes. The size of the viral genome can vary as much as 50-fold among different species of the group. The host range covers the range of living organisms. Eubacteria, Archaeans, single-celled eurkaryotic organisms, and various animals are infected by one or another Group I virus. The group has been subclassed based on shared morphologic properties, six of these subclasses contain human pathogens: Adenoviridae, Herpesviridae, Poxviridae, Papillomaviridae, Polyomaviridae, and Mimiviridae.

Most of the DNA-transforming viruses (i.e. DNA viruses that cause cancer) belong to Group I: Polyomaviruses, Adenoviruses, Papillomaviruses, and Herpesviruses (including Epstein–Barr virus). The one exception is Hepatitis B virus, which belongs to Group VII (Chapter 45). Unlike the retroviruses (Group VI, Chapter 44), which contain genes that are homologous with cancer-causing oncogenes, the DNA-transforming viruses do not contain oncogenes. The Group I DNA transforming viruses seem to cause cancer through a mechanism related to their ability to induce replication in their host cells.

Group I, dsDNA
  Herpesvirales
    Herpesviridae

*Epstein–Barr virus
*Herpes simplex virus type 1
*Herpes simplex virus type 2
*Herpes virus varicella-zoster virus
*Herpesvirus simiae, also known as B virus
*Human herpesvirus type 6, HHV6
*Human herpesvirus type 7, HHV7
*Human herpesvirus type 8, HHV8
*Cytomegalovirus

Members of Class Herpesviridae are commonly known as herpesviruses. These viruses produce acute disease characterized by lytic (i.e. cytopathic) effects in infected cells; and latent disease, characterized by recurrences of disease, sometimes spanning the life of the host. After cells are infected by virus particles, the viral genome migrates to the host nucleus, where replication and transcription of the viral genes occurs. During the latent phase, viruses may trigger a lytic phase, manifesting as clinical disease. The recurring disease may be clinically distinct from the initial infection (e.g. chicken pox, the initial varicella virus infection, is followed decades later by shingles). Some of the herpesviruses are DNA-transforming viruses.

The human herpesviruses are: Epstein–Barr virus, herpes simplex viruses, varicella virus, and human herpesviruses 6, 7, and 8, and cytomegalovirus.

Epstein–Barr virus infects almost all adults. Its persistence makes it one of the most prevalent human pathogens. It manifests acutely as mononucleosis, a pharyngitis accompanied by lymphocytosis (increased lymphocytes in the peripheral blood) and with morphologic alterations in infected lymphoctyes. Splenomegaly and hepatomegaly may occur. The generalized symptoms of the disease, particularly fatigue, may extend for months or longer, and some cases of mononucleosis recur. Epstein–Barr virus is a DNA-transforming virus and accounts for several cancers, including Hodgkin's lymphoma, Burkitt's lymphoma, nasopharyngeal carcinoma, and central nervous system lymphoma. A role for the virus in several autoimmune diseases has been suggested.

Herpes simplex type 1 causes cold sores, and herpes simplex types 2 causes genital herpes. Both diseases may recur after initial infection.

Herpes virus varicella-zoster causes chickenpox on first infection and herpes zoster, also known as shingles, on re-activation.

Herpesvirus simiae, also known as B virus, infects macaque monkeys, without causing severe disease. In rare circumstances, humans may become infected with this virus, from the monkey reservoir. Human infection typically results in a severe encephalopathy.

Human herpesvirus type 6 (HHV6) and type 7 (HHV7) produce exanthem subitum, also known as roseola infantum and sixth disease. Readers should not confuse sixth disease with fifth disease. Fifth disease, also known as erythema infectiosum and slapped face disease, is caused by Parvovirus

B19 (Chapter 40). These diseases take their names from an historical diagnostic dilemma faced by pediatricians, who regularly encountered six clinical syndromes of childhood rashes. Four of the childhood rashes had known etiologies. The fifth and sixth rashes, both caused by organisms that were not yet identified, were referred to as "fifth disease" and "sixth disease". Subsequently, the viral causes of these two diseases were discovered, but the numeric names held.

Human herpesvirus type 8 (HHV8) is a DNA-transforming virus that can cause Kaposi's sarcoma, primary effusion lymphoma, and some forms of Castleman's disease. Kaposi's sarcoma is a cancer characterized by focal proliferations of small blood vessels, occurring most often in the skin. Immune-suppressed patients (e.g. transplant recipients) who are carriers of the latent HHV8 virus, may develop Kaposi's sarcoma within a few months of immunosuppression. Interestingly, if immunosuppression is halted, the Kaposi's sarcoma may regress [124]. It is presumed that sustained viral replication is necessary for early tumor growth.

Cytomegalovirus infects about half of the world population, with most individuals suffering no ill-effects. Once infected, the virus usually persists for the life of the individual. In a minority of cases, particularly among immune-compromised individuals (e.g. organ transplant recipients and AIDS patients) and newborns, the virus may produce severe neurologic disease. The disease is known as cytomegalic inclusion body disease, and, as the name suggests, a large nuclear inclusion body characterizes actively infected cells. When the virus is transmitted transplacentally, by mothers infected during their pregnancy, the newborn may suffer developmental damage to the brain and other organs.

- Group I, dsDNA
  - Unassigned
    - Nonenveloped
      - Adenoviridae
        - *Human adenoviruses A through G
      - Papillomaviridae
        - *Human papillomavirus
      - Polyomaviridae
        - *BK polyomavirus
        - *JC polyomavirus
        - *Simian virus 40

Class Adenoviridae contains the human adenoviruses of which there are 57 types, with different clinical syndromes associated with specific subtypes of the virus. Most adenoviral diseases are either respiratory, conjunctival (i.e. viral conjunctivitis), or gastroenteritic. Infections may present clinically as tonsillitis (simulating strep throat), pharyngitis (croup), otitis media, pneumonia, meningoencephalitis, and hemorrhagic cystitis. Adenoviruses are commonly spread by aerosolized droplets, and are particularly stable in the external environment.

Human papillomaviruses cause skin warts, laryngeal warts, and genital warts. Warts are benign tumors composed of proliferating squamous cells. In some cases, these human papillomavirus-induced warts progress to become invasive squamous cell carcinomas.

Class Polyomaviridae contains several viruses that infect humans: BK polyomavirus, JC polyomavirus, and simian virus 40.

The BK polyomavirus rarely causes disease in infected patients, and the majority of humans carry the latent virus. Latency can shift to lytic infection after immunosuppression, producing a clinical nephropathy.

The JC polyomavirus persistently infects the majority of humans, but it is not associated with disease in otherwise healthy individuals. Rarely, in immune-compromised patients, JC polyomavirus may produce progressive multifocal leukoencephalopathy. The virus targets myelin-producing oligodendrocytes in the brain to produce areas of demyelination and necrosis.

Simian virus 40 (SV40) infects monkeys and humans, but there is no evidence at this time confirming a role in human disease.

- Group I, dsDNA
  - Unassigned
    - Nucleocytoplasmic large DNA viruses (NCLDV viruses)
      - Poxviridae
        - Orthopoxvirus
          - *Buffalopox virus
          - *Cowpox virus
          - *Monkeypox virus
          - *Vaccinia virus
          - *Variola major virus
          - *Variola minor virus
        - Parapoxvirus
          - *Orf
          - *Milker's nodule virus
        - Molluscipoxvirus
          - *Molluscum contagiosum virus
        - Yatapoxvirus
          - *Tanapoxvirus
          - *Yaba monkey tumor virus
          - *Yaba-like disease virus

Members of Class Orthopoxvirus produce disease characterized by pustules of the skin, and lymphadenopathy.

The smallpox virus is remarkable for its extremely narrow host range: humans only. The virus infects the skin and the mucosa of the upper respiratory tract, where it produces a pustular, weeping, rash. In the respiratory mucosa, the rash interferes with breathing. If the disease becomes hemorrhagic, the prognosis worsens. Smallpox is reputed to have killed about 300 million people in the twentieth century, prior to the widespread availability of an effective vaccine. Smallpox has been referred to as the greatest killer in human history. Its mortality rate was 30–35%, which is significantly less than some of the hemorrhagic viruses (e.g. Ebola virus, Group V, has a mortality rate of nearly 90%). No doubt, death rates climbed because the disease was easily communicable: via aerosols, fomites, bodily fluids, or direct contact with patients with rash.

Currently, smallpox has the distinction of being the only infection of humans which has been declared "eradicated". It should be noted that there are a few closed societies for which the status of smallpox in the population is unknown (e.g. North Korea). Aside from the remarkable success of vaccination, eradication was no doubt made possible because smallpox has no (non-human) animal reservoir. At this time, vaccination is not routinely performed and is reserved primarily as an antiterror measure, for personnel entering a zone where there is a bioweapons threat.

Variola minor is a virus closely related to Variola major that produces a milder disease. This disease is known by various names including alastrim, cottonpox, milkpox, whitepox, and Cuban itch. Infection with Variola minor is thought to produce cross-resistance to Variola major (and vice versa).

Vaccinia virus is the laboratory-grown poxvirus, of obscure heritage, that does not precisely correspond to known viruses that reside outside the laboratory or clinic (i.e. not quite cowpox, not quite variola). Vaccinations with vaccinia virus have been known to produce, in rare cases, a variety of clinical disorders, ranging from vaccinia (localized pustular eruptions), to generalized vaccinia, to progressive vaccinia, to vaccinia gangrenosum, to vaccinia necrosum. Other conditions associated with vaccinations include eczema vaccinatum and post-vaccinial encephalitis.

Smallpox vaccination, aside from eradicating mankind's greatest killer, may have heretofore unrecognized public health value. The number of currently known pathogenic organisms, their variant subtypes, their ability to mutate, and the emergence of newly encountered pathogens, make it impossible to develop vaccines for every organism that infects humans. Consequently, vaccine experts are searching for vaccines that confer immunity, partial or full, for several different pathogens or for several variants of a single pathogen [125]. An interesting development in this field is that the smallpox vaccine may confer limited protection against HIV infection. Both viruses enhance their infectivity by exploiting a receptor, CCR5, on the surface of white blood cells. This shared mode of infection may contribute to the cross-protection against HIV that seems to come from smallpox vaccine. It has been suggested that the emergence of HIV in the 1980s may have resulted, in part, from the cessation of smallpox vaccinations in the late 1970s [126].

Buffalopox, cowpox, and monkeypox produce diseases in animal reservoirs and rarely infect humans. Human infections occur from close contact with infected animals and manifest much like smallpox, but milder.

Members of Class Parapoxvirus infect vertebrates, particularly sheep, goats, cattle, and red squirrels. Orf virus causes "sore mouth" or "scabby mouth" disease of sheep and goats. Humans, though rarely infected, may develop painful hand sores. A similar condition can occur in humans who handle the udders of cows infected with Milker's nodule virus.

Class Molluscipoxvirus contains one species infectious in humans, *Molluscum contagiosum* virus. *Molluscum contagiosum* is an eruption of wart-like skin lesions that are easily diagnosed on histological examination by their distinctive cellular inclusions (so-called molluscum bodies). There are no known animal reservoirs. Infection is spread from human to human. Treatment is not always necessary, as individual lesions will regress within two months. However, auto-inoculation of the virus may produce new skin lesions, thus prolonging the disease.

Members of Class Yatapoxvirus infect primates in equatorial Africa. Infections can spread to humans by insect vectors. Tana poxvirus produces a pock-forming skin infection, with fever and lymphadenopathy in infected humans (like a mild form of smallpox). The Yaba monkey tumor virus produces histiocytomas in monkeys. Histiocytomas are proliferative lesions of fibrous tissue that yield tumor-like nodules. These virally induced histiocytomas in monkeys grow rapidly following infection, and then regress over the ensuing month [127]. Yaba monkey tumor virus and Yaba-like disease virus, like all members of Class Yatapoxvirus, are considered potential human pathogens.

Group I, dsDNA
- Unassigned
  - Nucleocytoplasmic large DNA viruses (NCLDV viruses)
    - Mimiviridae
      - Mimivirus
        - *Acanthamoeba polyphaga mimivirus (pneumonia)

Class Mimiviridae, discovered in 1992, occupies a niche that seems to span the biological gulf separating living organisms from viruses. Members of Class Mimiviridae are complex, larger than some bacteria, with enormous genomes (by viral standards), exceeding a million base pairs and encoding upwards of 1000 proteins. The large size and complexity of Class Mimiviridae exemplifies the advantage of a double-stranded DNA genome. DNA is much more chemically stable than RNA, and can be faithfully replicated, even when its length exceeds a billion base pairs. A double-stranded DNA genome can be protected by DNA repair enzymes, and by external modifications to the DNA structure. Class Megaviridae is a newly reported (October, 2011) class of viruses, related to Class Mimiviridae, but larger [128].

Biologically, the life of a mimivirus is not very different from that of obligate intracellular bacteria (e.g. Rickettsia). The discovery of Class Mimiviridae inspires biologists to reconsider the "non-living" status relegated to viruses and compels taxonomists to examine the placement of viruses within the phylogenetic development of prokaryotic and eukaryotic organisms.

Acanthamoeba polyphaga mimivirus is a possible human pathogen. Some patients with pneumonia have been shown to have antibodies against the virus [129].

Though Myxoma virus is not a human pathogen, it seems appropriate to include some mention of this member of Class Poxviridae, due to the role humans have played in its history. Myxoma virus produces a fatal disease, myxomatosis, in rabbits. The disease is characterized by the rapid appearance of skin tumors (myxomas), followed by severe conjunctivitis, systemic symptoms, and fulminant pneumonia. Death usually occurs 2–14 days after infection. In 1952, a French virologist, hoping to reduce the rabbit population on his private estate, inoculated a few rabbits with Myxoma virus. The results were much more than he had bargained for. Within two years, 90% of the rabbit population of France had succumbed to myxomatosis.

European rabbits, introduced to Australia in the nineteenth century, became feral and multiplied. By 1950 the rabbit population of Australia was about 3 billion. Seizing upon the Myxoma virus as a solution to rabbit overpopulation, the Australians launched a Myxoma virus inoculation program. In less than ten years, the Australian rabbit population was reduced by 95% [130]. Nearly 3 billion rabbits died, a number very close to the number of humans living on the planet in the mid-1950s. This plague on rabbits was unleashed by a committee of humans who decided that it was proper to use a lethal rabbit virus as a biological weapon. Without commenting on the moral implications of animal eradication efforts, it is worth noting that rabbits are not the only mammals that can be exterminated by a pathogenic virus. Humans should take heed.

Infectious species:

Adenoviridae: (type-specific clinical syndromes including respiratory, including pharyngitic and pneumonic, conjunctival, gastroenteritic, or bacteremic infections)

Human adenovirus A, types 12, 18, 31
Human adenovirus B, types 3, 7, 11, 14, 16, 21, 34, 35, 50, 55
Human adenovirus C, types 1, 2, 5, 6, 57
Human adenovirus D, types 8, 9, 10, 13, 15, 17, 19, 20, 22, 23, 24, 25, 26, 27, 28, 29, 30, 32, 33, 36, 37, 38, 39, 42, 43, 44, 45, 46, 47, 48, 49, 51, 53, 54, 56
Human adenovirus E, type 4
Human adenovirus F, types 40, 41
Human adenovirus G, type 52

Herpesviridae:

- Epstein–Barr virus (infectious mononucleosis, Hodgkin lymphoma, Burkitt lymphoma, nasopharyngeal carcinoma, central nervous system lymphoma, various autoimmune diseases)
- Herpes simplex types 1 (cold sores)
- Herpes simplex types 2 (genital herpes)
- Herpes virus varicella-zoster (chickenpox on first infection, herpes zoster or shingles on re-activation)
- Herpesvirus simiae, also known as B virus (encephalopathy)
- Human herpesvirus type 6, HHV6 (exanthem subitum, roseola infantum, sixth disease)
- Human herpesvirus type 7, HHV7 (exanthem subitum, roseola infantum, sixth disease; suggested but disputed cause of pityriasis rosea)
- Human herpesvirus type 8, HHV8 (Kaposi sarcoma, primary effusion lymphoma, Castleman's disease)
- Cytomegalovirus, also known as Human herpesvirus 5 (cytomegalic inclusion body disease)

Poxviridae:

- Variola major (smallpox)
- Vaccinia virus (vaccinia, generalized vaccinia, progressive vaccinia, vaccinia gangrenosum, vaccinia necrosum, eczema vaccinatum, post-vaccinial encephalitis)
- Variola minor (alastrim, cottonpox, milkpox, whitepox, and Cuban itch)
- Buffalopoxvirus (Buffalopox)
- Cowpox virus (Cowpox)
- Milker's nodes virus (Milker's nodes)
- Molluscum contagiosum virus (Molluscum contagiosum)
- Monkeypox virus (Monkeypox)
- Orf virus (Orf)
- Tana (mild pock-forming skin infection) [127]
- Yaba monkey tumor virus (regressing histiocytoma) [127]

Papillomaviridae:

- Human papillomavirus (warts, genital warts, laryngeal papillomas, squamous carcinoma)

Polyomaviridae:

- BK polyomavirus or BK virus (nephropathy in immune-compromised individuals)
- JC polyomavirus or JC virus (progressive mulifocal leukoencephalopathy in immune-compromised individuals)
- Simian virus 40 or SV40 virus (highly controversial potential cause of human cancer)

Mimiviridae:

- Acanthamoeba polyphaga mimivirus (pneumonia)

Chapter 40

Group II Viruses: Single-Stranded (+)Sense DNA

"Everything should be made as simple as possible, but not simpler."

Albert Einstein

Group I, dsDNA (Chapter 39)
Group II, ssDNA (Chapter 40)
  Parvoviridae
    *Bocavirus
    *Human parvovirus B19
Group III, dsRNA (Chapter 41)
Group IV (+)ssRNA (Chapter 42)
Group V (−)ssRNA (Chapter 43)
Group VI, ssRNA-RT (Chapter 44)
Group VII, dsDNA-RT (Chapter 45)
Prions (Chapter 46)

Group II, the single-stranded DNA viruses, contains only one family of viruses that are pathogenic in humans: Class Parvoviridae. Class Parvoviridae contains environmentally resistant viruses that infect a wide range of animals. Parvoviruses are the smallest viruses currently known.

Group II, ssDNA
  Parvoviridae
    *Bocavirus
    *Human parvovirus B19

Parvovirus B19 is the agent that causes fifth disease, so named because the diseases was the fifth type of common childhood exanthem among six rashes listed in textbooks. Until the 1980s, the cause of this fifth listed exanthem was unknown; so it came to be known as "fifth disease". Other names for fifth disease are erythema infectiosum and slapped face disease. Human

J.J. Berman: Taxonomic Guide to Infectious Diseases. DOI: http://dx.doi.org/10.1016/B978-0-12-415895-5.00040-4

herpesviruses 6 and 7 (Chapter 39) cause the sixth childhood rash, "sixth disease". The rash of fifth disease results from an immune response of the host to the virus particles. Essentially, fifth disease is an allergic phenomenon, and not the direct, cytopathic effect of the virus.

Infection by parvovirus occurs from contact (usually via respiratory droplets) with actively infected hosts. The virus is known to infect humans and dogs. Serologic evidence indicates that at least half of the human population has been infected with parvovirus B19.

Members of Class Parvoviridae characteristically infect rapidly dividing host cells, using host processes to support their own replication. The target cells for parvovirus B19 are the dividing precursor erythroid cells. Another name for parvovirus B19 is erythrovirus B19, indicating the target cell for the virus. In the active stage of infection, huge amounts of virus are produced. Death or dysfunction of the target hematopoietic (blood precursor) cells can lead to a transient pancytopenia (i.e. anemia of all blood cell lineages). In rare cases, aplastic anemia may occur, in which most of the precursor erythroid cells are destroyed, leading to a massive decline in circulating mature forms. When aplastic anemia occurs, it usually occurs in individuals who have a concurrent condition that requires an excessive production of blood cells to maintain the normal blood profile of mature cells. These conditions include: autoimmune hemolytic anemia, sickle cell anemia, and inherited blood dyscrasias that increase the fragility of red blood cells or that decrease the life-span of red blood cells. Basically, a co-infection with parvovirus B19 is the last straw for bone marrows that are barely keeping pace with the body's demand for erythrocytes.

The intense viremia that occurs in parvovirus B19 infection, and the small size of parvovirus particles, may predispose to cross-placental transmission occurring in some cases of infection in pregnant women. Though rare, parvovirus may cause miscarriage or hydrops fetalis (fluid accumulation in the fetus) with anemia.

Bocavirus has been associated with some cases of respiratory disease and diarrhea in young children. Though it is rarely detected in healthy persons, there is indication that it can occur in up to 9% of pediatric patients hospitalized with lower respiratory infections [131]. Bocavirus should not be confused with Bocas virus, a type of Coronavirus (Group IV, Chapter 42).

SEN virus (SEN-V) is a newly discovered single-stranded non-enveloped DNA virus that has been found in the blood of donors and recipients of transfusion blood [132]. In addition, another Group II virus, TT virus, also known as transfusion transmitted virus or Torque teno virus, has been isolated from transfusion blood. TT virus is currently a suspected hepatitis virus. At this time, the pathogenicity of both SEN-V and TT viruses are in doubt; hence neither virus is included in the list of Group II virus pathogens.

Infectious species:

Bocavirus (respiratory disease and diarrhea in children)
Human parvovirus B19, alternately known as erythrovirus B19 (fifth disease, erythema infectiosum, slapped face disease, transient hemolytic anemia, aplastic anemia)

Chapter 41

Group III Viruses: Double-Stranded RNA

We are just "a volume of diseases bound together."

John Donne

Group I, dsDNA (Chapter 39)
Group II, ssDNA (Chapter 40)
Group III, dsRNA (Chapter 41)
  Reoviridae
    *Rotavirus
    *Coltivirus
    *Orbivirus
Group IV (+)ssRNA (Chapter 42)
Group V (−)ssRNA (Chapter 43)
Group VI, ssRNA-RT (Chapter 44)
Group VII, dsDNA-RT (Chapter 45)
Prions (Chapter 46)

The Group III viruses have a double-stranded RNA genome. Replication of Group III viruses takes place exclusively in the cytoplasm, where the viral RNA codes for the proteins needed for viral replication. Viral proteins are synthesized using the host cell machinery (i.e. ribosomes).

Group III contains six major classes, only one of which contains organisms that are infectious in humans: Class Reoviridae. The name derives from "Respiratory enteric orphan" viruses. The term "orphan", when applied to a virus, indicates that no known diseases are associated with the virus. This is no longer the case for the Reoviruses.

Group III, dsRNA
  Reoviridae
    *Rotavirus
    *Coltivirus
    *Orbivirus

J.J. Berman: Taxonomic Guide to Infectious Diseases. DOI: http://dx.doi.org/10.1016/B978-0-12-415895-5.00041-6

The most clinically significant species in Class Reoviridae is rotavirus. In 2004, rotavirus infections accounted for about a half million deaths in young children, from severe diarrhea [5]. Most of the deaths occurred in developing countries. The death rate is expected to decline, due to the recent introduction of an apparently safe and effective vaccine [5]. Rotavirus, when observed with transmission electron microscopy, resembles a wagon wheel. It was formerly known as gastroenteritis virus type B. It is passed from human to human by fecal–oral route.

Aside from rotavirus, there are three genera that infect humans, two of which produce disease. The Orothoreoviruses infect vertebrates, including humans, but no disease has been linked to the infections. The two disease-producing infectious genera are Coltivirus and Orbivirus.

Colorado tick fever is endemic in the Rocky Mountains (in contradistinction to Rocky Mountain spotted fever, a rickettsial infection, discussed in Chapter 5, that has no restricted affinity for the Rocky Mountains). Coltivirus takes its name from the disease (i.e., COLorado TIck fever VIRUS). As the name suggests, Colorado tick fever is carried by a tick (in this case, *Dermacentor andersoni*) and produces a fever. The fever is often accompanied by myalgia, headache, and photophobia. In a small percentage of children with Colorado tick fever, encephalitis may follow.

Orbiviruses have been implicated in several rather obscure fever-associated conditions: Kemerovo fever, found in Western Siberia and transmitted by ticks; Orungo fever, found in Central African and transmitted by mosquitoes; and Changuinola fever, found in northern South America and transmitted by sand flies of Class Phlebotomus.

Infectious species:

Human rotavirus (gastroenteritis, diarrhea)
Coltivirus (Colorado tick fever)
Orbivirus species (Colorado tick fever, Kemerovo fever, Orungo fever, Changuinola fever, and other midge-, mosquito-, or tick-borne geographically-confined viral syndromes)

Chapter 42

Group IV Viruses: Single-Stranded (+)Sense RNA

"Simplicity is the ultimate sophistication."

Leonardo da Vinci

Group I, dsDNA (Chapter 39)
Group II, ssDNA (Chapter 40)
Group III, dsRNA (Chapter 41)
Group IV (+)ssRNA (Chapter 42)
  Nidovirales
    Coronaviridae
      *SARS virus
      *Torovirus species
  Hepeviridae:
    *Hepatitis E
  Caliciviridae
    *Norovirus, formerly Norwalk virus
    *Sapporo virus
  Togaviridae
    Alphavirus viral diseases
      *Chikungunya
      *Eastern equine encephalomyelitis virus
      *Getah virus
      *Mayaro virus
      *Mucambo virus
      *O'nyong'nyong virus
      *Ross river virus
      *Barmah forest virus
      *Sagiyama virus
      *Semliki forest virus
      *Sindbis virus
      *Tonate virus

J.J. Berman: Taxonomic Guide to Infectious Diseases. DOI: http://dx.doi.org/10.1016/B978-0-12-415895-5.00042-8

*Venezuelan equine encephalomyelitis virus
*Western equine encephalomyelitis virus
Rubivirus
*Rubella virus
Flaviviridae
Hepacivirus
*Hepatitis C
Flavivirus
*Dengue virus types 1–4
*Hepatitis G virus
*Japanese B encephalitis virus
*Murray Valley encephalitis virus
*Rocio virus
*Spondweni virus
*St Louis encephalitis
*Wesselsbron
*West Nile virus (West Nile fever)
*Yellow fever virus (yellow fever)
Tick-borne virus group
*Absettarov
*Hanzalova
*Hypr
*Kumlinge
*Kyasanur forest disease
*Louping ill (tick-borne encephalitis)
*Negishi
*Omsk
*Powassan (tick-borne encephalitis)
*Langat (tick-borne encephalitis)
*Russian spring summer encephalitis
Hepatitis G virus group
*Hepatitis G virus
Picornaviridae
Enterovirus
*Coxsackievirus
*Echovirus
*Poliovirus
*Enterovirus 68–109
*Rhinovirus A and B
Hepatovirus
*Hepatitis A, alternately human enterovirus type 72
Astroviridae
*Astrovirus species
Group V (−)ssRNA (Chapter 43)

Group VI, ssRNA-RT (Chapter 44)
Group VII, dsDNA-RT (Chapter 45)
Prions (Chapter 46)

The Group IV viruses have a positive sense RNA genome. Positive sense RNA can be translated directly into protein, without a DNA intermediate and without creating a complementary RNA strand. To replicate its genome, though, a complementary RNA strand is required. The positive RNA strand serves as a template for an RNA-dependent RNA polymerase, yielding a complementary RNA strand, to form a dimer with the template strand. The double-stranded RNA subsequently serves as the template for a new positive sense genome.

The positive strand RNA genome is independently infectious, for most Group IV viruses. This means that in the absence of a capsid, envelope, or enclosed proteins, the RNA molecule, when inserted into a cell, is capable of using host cell machinery to construct additional viruses. Such viruses can be extremely small. In an experiment conducted in the late 1960s, Sol Spiegelman and his coworkers developed a method by which smaller and smaller viral RNA molecules could be isolated that were capable of replicating if provided with a specific RNA-dependent RNA polymerase and substrate nucleotides. A minimalist infectious viral genome was eventually selected that was only 220 nucleotides (bases) in length [133].

As discussed in the overview (Chapter 38), we know very little about the phylogenetic relationship among viruses. Consequently, we subclassify the Group IV viruses based on structural similarities: symmetry of capsid, presence of absence of a viral envelope, size. There are six subclasses of the Group IV single-stranded positive-sense RNA viruses: Picornaviridae, Togaviridae, Coronaviridae, Hepeviridae, Caliciviridae, Flaviviridae, and Astroviridae. As expected, within each class, viruses share structural similarities; but there are no properties, other than the defining property of a (+)sense single-stranded RNA genome, that extends to all six classes of the Group IV viruses. For example, some classes have envelopes (i.e. Flaviviridae, Togaviridae, Coronaviridae), and others do not (i.e. Caliciviridae, Picornaviridae, Hepeviridae, and Astroviridae).

Group IV (+)ssRNA
  Nidovirales
    Coronaviridae
      *SARS virus
      Torovirinae
        *Human torovirus

Members of Class Nidovirales do not package polymerase in the viral particle. The genome is read directly using host enzymes. Class Nidovirales contains one subclass with viruses that infect humans: Class Coronaviridae. The coronaviruses are characterized by glycoprotein spikes (peplomers) that protrude from the envelope that encloses the round nucleocapsid. The arrangement of peplomers resembles a corona (hence, coronavirus).

SARS (severe acute respiratory syndrome) virus produces a flu-like illness, and is spread by close human contact. The earliest outbreak of SARS occurred in Southeast Asia, in 2002. Soon thereafter, cases occurred in distant locations, including Toronto, Canada and Bangalore, India. The worldwide response to SARS was possibly the most intensive public health effort ever launched to stem an epidemic of a new, and potentially fatal, viral disease. By 2004, China, the epicenter of the fledgling epidemic, was declared SARS-free.

Toroviruses infect a variety of mammals. Human torovirus is a rare cause of enteritis.

Group IV (+)ssRNA
  Hepeviridae
    *Hepatitis E

Class Hepeviridae is a tentative class, with just one genus and one species, the hepatitis E virus. It is possible that the hepatitis E virus will be re-assigned to an existing subclass of Group IV or will be assigned to a newly named subclass. Hepatitis E is spread from human to human by a fecal–oral route or by food contaminated with feces of infected mammals. The resulting hepatitis has a clinical course similar to that seen with Hepatitis A (i.e., self-limited disease, not leading to cirrhosis). Pregnant women infected with Hepatitis E are at increased risk for developing fulminant hepatitis, which may be fatal.

Hepevirus should not be confused with the orthographically similar "herpesvirus". Also, readers should not confuse Class Hepeviridae (Hepatitis E virus) with Class Hepacivirus, a subclass of Class Flaviviridae that contains the Hepatitis C virus. Neither of these Group IV subclasses should be confused with Class Hepadnaviridae (Group VII, Chapter 45).

Group IV (+)ssRNA
  Caliciviridae
    *Norovirus, formerly Norwalk virus
    *Sapporo virus

Class Caliciviridae contains small non-enveloped viruses, 35–40 nm in diameter, just a tad larger than members of Class Picornaviridae. They take their name from the Latin root, calyx, meaning goblet, referring to the cup-shaped capsid depressions. Members of Class Caliciviridae cause acute gastroenteritis in humans.

Group IV (+)ssRNA
  Togaviridae
    Alphavirus diseases
      *Chikungunya
      *Eastern equine encephalomyelitis virus

*Getah virus
*Mayaro virus
*Mucambo virus
*O'nyong'nyong virus
*Ross river virus
*Barmah forest virus
*Sagiyama virus
*Semliki forest virus
*Sindbis virus
*Tonate virus
*Venezuelan equine encephalomyelitis virus
*Western equine encephalomyelitis virus
Rubivirus
*Rubella virus

Class Togaviridae is named for its distinctive coat (the "toga"). Togaviruses have a genome approximately 12 000 kilobases in length, somewhat larger than the genome of Class Picornaviridae. Togaviruses live in the cytoplasm of their host cells, where viral replication and gene expression take place. Class Togaviridae contains two subclasses: Class Alphavirus and Class Rubivirus. All the members of Class Alphavirus are arboviruses (arthropod-borne viruses) spread by mosquitoes (primarily) or ticks. Class Rubivirus contains only one species that is infective in humans: Rubella virus, the cause of German measles. Rubella is spread directly from person to person, without an insect vector. Readers should not confuse Rubella virus with the measles virus, Rubeola. Rubeola virus is a paramyxovirus (Group V, Chapter 43), unrelated to Rubella virus.

A few of the alphaviruses typify the group. Chikungunya is a disease that produces a clinical syndrome similar to that seen with Dengue virus (Class Flaviviridae), Ross river virus and Barmah forest virus; namely, an acute febrile phase followed by a prolonged arthralgic phase. Chikungunya fever is spread by the Aedes mosquito, and the reservoir is primarily human (i.e. transmission is human to mosquito to human). In recent years, the incidence of Chikungunya has recently increased in Asia and Africa and is now an emerging disease in Europe [76].

Ross river virus and Barmah forest virus produce clinically and geographically indistinguishable diseases, sometimes referred to collectively as epidemic polyarthritis. The diseases are spread by various species of mosquito, and both are endemic to Australasia. They produce an acute influenza-like illness, followed by arthralgia. Joint pains may persist for many months. The reservoir for both viruses seem to be, primarily, marsupials.

Group IV (+)ssRNA
Flaviviridae
Hepacivirus
*Hepatitis C

Flavivirus
*Dengue virus types 1–4
*Japanese B encephalitis virus
*Murray Valley encephalitis virus
*Rocio virus
*Spondweni virus
*St Louis encephalitis
*Wesselsbron
*West Nile virus
*Yellow fever virus
Tick-borne virus group
*Absettarov virus
*Hanzalova virus
*Hypr virus
*Kumlinge virus
*Kyasanur forest disease virus
*Louping ill (tick-borne encephalitis)
*Negishi virus
*Omsk virus
*Powassan (tick-borne encephalitis)
*Langat (tick-borne encephalitis)
*Russian spring summer encephalitis
Hepatitis G virus group
*Hepatitis G virus

The members of Class Flaviviridae are enveloped, spherical, and have a diameter of about 50 nm. Most members of Class Flaviviridae are arthropod-borne, being transmitted by a tick (Class Chelicerata, Chapter 29) or a mosquito (Class Hexapoda, Chapter 30). The subclasses of Class Flaviviridae that contain infectious viruses in humans are: Hepacivirus, Flavivirus, tick-borne virus group, and hepatitis G virus group.

Hepatitis C virus is the only member of Class Hepacivirus. Hepatitis C is a common cause of hepatitis and chronic liver disease. It is spread from person to person by sexual transmission or by contact with infected blood, or blood products, and can be spread by contaminated needles. It can be transmitted to infants born to infected mothers. People who develop hepatitis from this virus often develop chronic infection of the liver, that varies from person to person in intensity and in the likelihood of progressing to cirrhosis. Over one percent of the USA population is infected with hepatitis C.

Class Flavivirus (from the Latin "flavus," meaning yellow), is a subclass of Class Flaviviridae, both named for the yellow (jaundiced) skin resulting from infection with its most notorious species, Yellow fever virus. The flaviviruses includes some of the most common and deadly viruses on earth, led by yellow fever virus and Dengue fever virus. Among the flaviviruses are numerous encephalitis-producing viruses, that have specific geographic distributions: Japanese encephalitis virus (mosquito-borne), Murray Valley encephalitis virus (mosquito-borne), St. Louis encephalitis virus (mosquito-borne), West Nile encephalitis virus (mosquito-borne), and a host of viruses collectively known as tick-borne encephalitis viruses.

Yellow fever virus seems to have originated in Africa and spread to other continents in the mid-seventeenth century. It is responsible for hundreds of thousands of deaths in North America alone. The disease is carried by primates, including humans, and transmitted from person to person by the bite of a mosquito (*Aedes aegypti*). It produces hepatitis and hemorrhaging (hence, it is included among the hepatitis viruses and the hemorrhagic fever viruses).

Yellow fever virus is associated with an impressive number of medical breakthroughs, being the first disease demonstrated to be transmitted by an arthropod, among the first disease shown to be caused by a virus, and among the first infections controlled by a live vaccine. Effective methods of yellow fever prevention, through the eradication of the *Aedes aegypti* were developed in the 1890s, and an effective vaccine was developed in the 1930s. Today, there are about 200 000 cases of yellow fever, worldwide, with about 30 000 deaths [134]. Most infections occur in Africa.

While the incidence of yellow fever has diminished over the past century, the incidence of dengue fever is increasing. Dengue, like yellow fever, is transmitted primarily by the *Aedes aegypti* mosquito. More than 50 million dengue virus infections occur each year, worldwide. Most infections are asymptomatic or cause only mild disease, but a minority of infections are severe and may cause death. Typical cases exhibit fever and intense pain in muscles and joints (hence the alternate name of the disease, breakbone fever). Fevers can come and go. Capillary permeability is a common feature of the disease, and this may result in petechiae, the egress of fluid from the vascular compartment, shock (so-called Dengue shock syndrome), and hemorrhage (hemorrhagic syndrome). Severe cases of dengue, if untreated, may have a fatality rate approaching 20%. Like yellow fever, dengue is included in the group of hemorrhagic viruses.

The hepatitis G virus group, in Class Flaviviridae, contains only one species, hepatitis G virus, which had been traditionally included among the named hepatitis viruses. Hepatitis G is now considered to be an "orphan virus" (i.e., a virus that has no associated disease). The hepatitis G virus is found in a small percentage of donated blood units.

Group IV (+)ssRNA
- Picornaviridae
  - Enterovirus
    - *Coxsackievirus
    - *Echovirus

*Poliovirus
*Enterovirus 68–109
*Rhinovirus A and B
Hepatovirus
*Hepatitis A, alternately human enterovirus type 72

Members of Class Picornaviridae have a small RNA genome, as small as 7 kilobases (i.e., 7000 nucleotides) in length. The picornaviruses include two subclasses: Enterovirus and Hepatovirus.

Members of Class Enterovirus are among the most prevalent human pathogens. In active infections, the virus can often be recovered from feces and respiratory secretions. Poliovirus, spread by contaminated fecal material, produces a paralytic syndrome characterized by inflammation and destruction of the anterior horn cells of the spinal cord. Many poliovirus infections do not result in disease, but disease-free infected individuals are carriers, transiently producing infective virus. Aside from poliovirus, other enteroviruses may display neurotropism, producing aseptic meningitis and flacid paralysis [135].

There are a huge number of serotypes in Class Enterovirus, spread among the Coxsackievirus, Echovirus, and Enterovirus viruses. They produce non-specific flu-like illnesses, and various strains produce distinctive syndromes such as hand-foot-mouth disease, herpangina, and hemorrhagic conjunctivitis, and Bornholm disease (epidemic pleurodynia). Enterovirus infections are common pediatric maladies. Infections in newborns can be severe, with hepatitis, encephalitis, and sepsis.

Class Enterovirus also includes the rhinoviruses, which contains more than 100 variant strains. The rhinoviruses account for most instances of the common cold.

Class Hepatovirus contains hepatitis A, a cause of hepatitis. As you would expect from a member of Class Enterovirus, hepatitis A is typically spread by the fecal–oral route. The resulting hepatitis is acute, and generally subsides without sequelae. Unlike hepatitis B and C, hepatitis A does not progress to chronic hepatitis and cirrhosis.

Group IV (+)ssRNA
Astroviridae
*Astrovirus species

Members of Class Astroviridae, like those of Class Picornaviridae and Class Caliciviridae, lack an envelope. The class contains one species that is pathogenic in humans: Astrovirus. Astrovirus causes enteritis. Infections are especially common in children, accounting for more than 5% of enteritis cases in the pediatric age group.

Infectious species:

Enterovirus
Coxsackievirus (flu-like illness, hand-food-mouth disease, herpangina, hemorrhagic conjunctivitis, aseptic meningitis; in newborns, myocarditis, meningoencephalitis, hepatitis)

Echovirus, Enteric Cytopathic Human Orphan virus (flu-like illness, aseptic meningitis; in newborns, severe myocarditis hepatitis, and systemic infection)
Poliovirus (polio)
Enterovirus 68–109 (flu-like illnesses and other syndromes associated with Coxsackievirus and Echovirus)
Rhinovirus (common cold)
Hepatovirus
Hepatitis A (hepatitis A)
Alphavirus viral diseases
Chikungunya virus, a member of Semliki Forest virus complex (Chikungunya fever, rash, arthritis)
Eastern equine encephalomyelitis virus (encephalitis)
Getah virus (asymptomatic in humans)
Mayaro virus, a member of Semliki Forest virus complex (rash, arthritis)
Mucambo virus (encephalitis)
O'nyong'nyong virus, a member of Semliki Forest virus complex (rash, arthritis)
Ross river virus, a member of Semliki Forest virus complex (epidemic polyarthritis)
Barmah forest virus (epidemic polyarthritis)
Sagiyama virus, a subtype of Ross River virus (asymptomatic in humans)
Semliki forest virus, a member of Semliki Forest virus complex (rash, arthritis)
Sindbis virus (Sindbis fever, rash, arthritis)
Tonate virus (encephalitis)
Venezuelan equine encephalomyelitis virus (encephalitis, often causing, in humans, a flu-like illness with high fever and headache)
Western equine encephalomyelitis virus (Western equine encephalomyelitis)
Rubivirus
Rubella virus (German measles)
Coronaviridae
SARS virus (severe acute respiratory syndrome)
Torovirus species (gastroenteritis)
Hepeviridae
Hepatitis E (hepatitis)
Caliciviridae
Norovirus, formerly Norwalk virus (epidemic gastroenteritis)
Sapporo virus (mild gastroenteritis in children)
Flaviviridae
Hepacivirus
Hepatitis C (hepatitis C)
Flavivirus

Dengue virus types 1–4 (dengue fever; severe form is called dengue hemorrhagic fever)
Japanese B encephalitis virus (encephalitis, with less than 1% of human infections leading to disease)
Murray Valley encephalitis virus (Murray Valley encephalitis, formerly known as Australian encephalitis)
Rocio virus (meningoencephalitis)
Spondweni virus (acute febrile illness)
St Louis encephalitis (encephalitis)
Wesselsbron virus (fever, flu-like illness, most infections are subclinical)
West Nile virus (West Nile fever). *Aedes albopictus* is a vector for West Nile
Yellow fever virus (yellow fever). The yellow fever mosquito is *Aedes aegypti*

Tick-borne virus group
Absettarov virus (encephalitis)
Hanzalova virus (encephalitis)
Hypr virus (encephalitis)
Kumlinge virus (fever and encephalitis)
Kyasanur forest disease virus (hemorrhagic fever)
Louping ill virus (tick-borne encephalitis)
Negishi virus (encephalitis)
Omsk virus (hemorrhagic fever)
Powassan virus (tick-borne encephalitis)
Langat virus (tick-borne encephalitis)
Russian spring summer encephalitis virus (encephalitis)

Hepatitis G virus group
Hepatitis G virus, alternately GB virus CF ("orphan virus" not associated with any human disease)

Astroviridae
Astrovirus species (gastroenteritis)

Chapter 43

Group V Viruses: Single Stranded (−)Sense RNA

"An inefficient virus kills its host. A clever virus stays with it."

James Lovelock

Group I, dsDNA (Chapter 39)
Group II, ssDNA (Chapter 40)
Group III, dsRNA (Chapter 41)
Group IV (+)ssRNA (Chapter 42)
Group V (−)ssRNA (Chapter 43)
  Mononegavirales (nonsegmented)
    Paramyxoviridae
      Henipavirus
        *Hendra virus
        *Nipah virus
      Rubulavirus
        *Mumps virus
        *Parainfluenza types 2, 4a and 4b
      Morbillivirus
        *Measels virus, also known as Rubeola or Morbilli virus
      Avulavirus
        *Newcastle disease virus
      *Metapneumovirus
      Pneumovirus
        *Respiratory syncytial virus, RSV
      *Parainfluenza Types 1 to 4
    Rhabdoviridae
      Lyssavirus
        *Duvenhage
        *Rabies virus

J.J. Berman: Taxonomic Guide to Infectious Diseases. DOI: http://dx.doi.org/10.1016/B978-0-12-415895-5.00043-X

Vesiculovirus
*Vesicular stomatitis
*Chandipura
Filoviridae
Marburgvirus
*Marburg virus
Ebolavirus
*Ebola Reston
*Ebola Siena
*Ebola Sudan
*Ebola Zaire
Bornaviridae
*Bornavirus
Unassigned Classes
Arenaviridae
Lassa virus complex
*Lassa virus
*Lujo virus
*Lymphocytic choriomeningitis virus
*Mobala virus
*Mopeia virus
Tacaribe-virus complex
*Amapari virus
*Chapare virus
*Flexal virus
*Guanarito virus
*Junin virus
*Latino virus
*Machupo virus
*Oliveros virus
*Parana virus
*Pichinde virus
*Pirital virus
*Sabia virus
*Tacaribe virus
*Tamiami virus
*Whitewater Arroyo virus
Bunyaviridae
*Akabane virus
*Bunyamwera virus
*California encephalitis virus
Hantavirus
*Hantavirus Belgrade virus
*Hantavirus Hantaan virus

*Hantavirus Prospect Hill virus
*Hantavirus Puumala virus
*Hantavirus Seoul virus
*Hantavirus Sin Nombre virus
Nairovirus
*Nairovirus Bhanja virus
*Nairovirus Crimean virus
*Nairovirus Hazara virus
*Oropouche virus
Phlebovirus
*Rift Valley Fever (RVF) virus
*Pappataci fever virus
Orthomyxoviridae
*Influenza, types A, B and C (viral influenza)
Thogotovirus
*Tick-borne orthomyxoviridae Dhori
*Tick-borne orthomyxoviridae Thogoto
Unassigned genus
Deltavirus
*Hepatitis delta virus
Group VI, ssRNA-RT (Chapter 44)
Group VII, dsDNA-RT (Chapter 45)
Prions (Chapter 46)

Single-stranded RNA viruses are grouped into viruses with a positive sense, negative sense, and ambisense.

Positive sense RNA viruses are those that have the same direction as human mRNA. Positive sense virus genes can be translated (to yield protein) with the same cellular machinery that translates mRNA. These are the Group IV viruses (Chapter 42).

Negative sense RNA is complementary to mRNA. Negative sense RNA must be converted to positive sense RNA or to DNA before it becomes biologically available for translation or replication, respectively.

The Group V viruses (Chapter 43) are single-stranded negative sense RNA viruses that use an RNA-dependent RNA polymerase, packaged within the virus particle, to produce positive sense RNA, within the host cell. The Group VI viruses (Chapter 44) are single-stranded negative sense RNA viruses that use an RNA-dependent DNA polymerase (so-called reverse transcriptase), packaged within the virus particle, to produce a complementary strand of DNA. The synthesized strand of DNA is subsequently used as a template to yield a double-stranded DNA molecule that contains the genetic information from the viral genome.

The ambisense single-strand RNA viruses contain at least one positive sense RNA segment admixed with negative sense RNA. Genomically, the ambisense viruses are a hybrid mixture of Group IV virus (positive sense single-stranded) and a negative sense virus. Nevertheless, transcription in ambisense viruses is coupled with translation of the viral genome to a complementary RNA strand, a process that is characteristic of the Group V viruses [136]. Consequently, the ambisense viruses are currently included in the "unassigned" subclasses of the Group V viruses.

Among the Group V RNA viruses pathogenic to humans, there is one assigned class of viruses, Class Monongavirales, which includes the following subclasses: Bornaviridae, Filoviridae, Paramyxoviridae, and Rhabdoviridae. The remaining Class V viruses belong to unassigned subclasses (i.e., with no named taxonomic superclass) or unassigned genera (i.e. belonging to no assigned class). The unassigned classes are: Arenaviridae, Bunyaviridae, and Orthomyxoviridae. Deltavirus is an unassigned genus.

The Group V viruses are numerous, and it would be unproductive to describe each virus in detail. Readers who are interested can link the viral names (listed at the end of the chapter), with web resources. The following discussion will include viral disorders that typify their class or that highlight recent findings that might have been omitted from previously published virology texts. A listing of the Group V viruses, along with their associated diseases and clinical conditions, arranged by subclass, is found at the end of the chapter.

Group V (−)ssRNA
- Mononegavirales (nonsegmented)
  - Paramyxoviridae
    - Henipavirus
      - *Hendra virus
      - *Nipah virus
    - Rubulavirus
      - *Mumps virus
      - *Parainfluenza types 2, 4a and 4b
    - Morbillivirus
      - *Measels virus, also known as Rubeola or Morbilli virus
    - Avulavirus
      - *Newcastle disease virus
    - *Metapneumovirus
    - Pneumovirus
      - *Respiratory syncytial virus, RSV
    - *Parainfluenza Types 1 to 4
  - Rhabdoviridae
    - Lyssavirus
      - *Duvenhage
      - *Rabies virus

Vesiculovirus
*Vesicular stomatitis
*Chandipura
Filoviridae
Marburgvirus
*Marburg virus
Ebolavirus
*Ebola Reston
*Ebola Siena
*Ebola Sudan
*Ebola Zaire
Bornaviridae
*Bornavirus

Class Paramyxoviridae includes the viruses that cause measles (Rubeola or Morbilli virus) and mumps. Measles remains one of the most fatal virus disease, in terms of deaths worldwide. In 2001, measles virus accounted for about 745 000 deaths. Class Paramyxoviridae also includes several viruses that account for many respiratory diseases, particularly in children: the parainfluenza viruses, respiratory syncytial virus, and (somewhat less severe, clinically) human metapneumovirus.

Class Rhabdoviridae includes Rabies virus, a species of lyssavirus (from Lyssa, the Greek goddess of madness).

Class Filoviridae includes Ebola virus and Marburg virus. Members of Class Filoviridae infect primates (including humans) and produce potentially fatal viral hemorrhagic fevers.

Class Bornaviridae contains one species that infects humans, Borna virus. This virus, which infects a variety of animals, is neurotropic, producing nervous system inflammation (e.g. meningoencephalitis), neurologic impairment (e.g. ataxia) and behavior disorders (e.g. excitation or depression) in infected animals.

The role of Bornavirus in humans is undetermined at this time, but Bornavirus infections have been implicated as a cause of mental illnesses in humans, including bipolar disorder. It is the only member of Class Mononegavirales that replicates inside the host nucleus. Bornavirus has recently gained scientific attention as the only non-retrovirus that has been shown to integrate (permanently) into the mammalian genome [115]. Inherited fragments of bornavirus have been found in various types of mammals, suggesting that bornavirus is not new [113].

Group V (−)ssRNA
Unassigned Classes
Arenaviridae
Lassa virus complex
*Lassa virus (Lassa fever)

*Lujo virus
*Lymphocytic choriomeningitis virus
*Mobala virus
*Mopeia virus
Tacaribe-virus complex
*Amapari virus
*Chapare virus
*Flexal virus
*Guanarito virus
*Junin virus
*Latino virus
*Machupo virus
*Oliveros virus
*Parana virus
*Pichinde virus
*Pirital virus
*Sabia virus
*Tacaribe virus
*Tamiami virus
*Whitewater Arroyo virus
Bunyaviridae
*Akabane virus
*Bunyamwera virus
*California encephalitis virus
Hantavirus
*Hantavirus Belgrade virus
*Hantavirus Hantaan virus
*Hantavirus Prospect Hill virus
*Hantavirus Puumala virus
*Hantavirus Seoul virus
*Hantavirus Sin Nombre virus
Nairovirus
*Nairovirus Bhanja virus
*Nairovirus Crimean virus
*Nairovirus Hazara virus
*Oropouche virus
Phlebovirus
*Rift Valley Fever (RVF) virus
*Pappataci fever virus
Orthomyxoviridae
*Influenza, types A, B and C (viral influenza)
Thogotovirus
*Tick-borne orthomyxoviridae Dhori
*Tick-borne orthomyxoviridae Thogoto

Along with the bunyaviruses, the viruses in Class Arenaviridae account for the majority of roboviruses (rodent borne viruses). In general, each arenavirus infects a specific species of rodent. Rodents occasionally transmit the virus to humans, either through aerosolized excreta (e.g., urine, feces), or by direct contact of infectious material with cuts and abrasions in human skin. Most of the diseases caused by bunyaviruses are either encephalitides, hemorrhagic fevers, or non-hemorrhagic fevers; the severity varying with the viral species and host resistance.

The arenaviruses have been separated by geographic locale into two groups: New World viruses and Old World viruses. Old World arenaviruses and their approximate locales are: Ippy virus (Central African Republic), Lassa virus (West Africa), Lymphocytic choriomeningitis virus (worldwide), Mobala virus (Central African Republic), and Mopeia virus (Mozambique).

The New World arenaviruses and their approximate locales are: Amapari virus (Brazil), Flexal virus (Brazil), Guanarito virus (Venezuela), Junin virus (Argentina), Latino virus (Bolivia), Machupo virus (Bolivia), Oliveros virus (Argentina), Parana virus (Paraguay), Pichinde virus (Columbia), Pirital virus (Venezuela), Sabia virus (Brazil), Tacaribe virus (Trinidad), Tamiami virus (Florida, USA), and Whitewater Arroyo virus (New Mexico, USA). Several of the arenaviruses isolated from humans have not, as yet, been associated with human disease; these species are omitted from the list of infectious diseases caused by arenaviruses (see below).

One member of Class Arenaviridae, Lassa virus, the cause of Lassa fever, should not be confused with Lyssa virus, a member of Class Rhabdoviridae and the cause of rabies.

Class Bunyaviridae along with Class Arenaviridae, account for the majority of Roboviruses (rodent-borne viruses). In addition, like the virus syndromes produced by members of Class Arenaviridae, the members of Class Bunyaviridae tend to produce fever syndromes, hemorrhagic fever syndromes, or meningoencephalitides.

Class Orthomyxoviridae includes Influenza virus (types A, B, and C) and the thogotoviruses. Seasonal influenza kills between a quarter million and a half million people worldwide, each year. In the USA seasonal influenza accounts for about 40 000 deaths annually. The global 1917–1918 influenza pandemic caused somewhere between 50 million and 100 million deaths.

The thogotoviruses (Dhori virus and Thogoto virus) infect ticks, and ticks transmit the infection to humans. Thogotoviruses produce fever and encephalitis.

Group V (−)ssRNA
- Unassigned Classes
  - Unassigned genus
    - Deltavirus
      - *Hepatitis delta virus

Genus Deltavirus is a genus in Group V that has not been assigned a viral class. The genus contains one species, the Hepatitis D virus. The Hepatitis D virus cannot replicate on its own. Replication requires the presence of Hepatitis B virus in the same host cell. Co-infections of Hepatitis B and Hepatitis D viruses (sometimes referred to as Labrea fever) have a more severe clinical course than infections with Hepatitis B alone (e.g., increased likelihood of developing liver failure, shortened time interval for initial infection to the development of cirrhosis, increased likelihood of developing liver cancer, and overall increased mortality rate).

Infectious species:

Paramyxoviridae:

- Hendra virus, a henipavirus transmitted by Pteropid fruit bats (pulmonary edema and hemorrhage, encephalitis)
- Nipah, a henipavirus transmitted by Pteropid fruit bats (encephalitis, relapse encephalitis)
- Human metapneumonovirus (mild respiratory illness in healthy individuals, occasionally severe respiratory illness in children, elderly, and immune-compromised individuals)
- Measles virus (measles, rubeola, morbilli)
- Mumps virus (mumps)
- Newcastle disease virus, transmitted by infected birds (conjunctivitis and flu-like illness)
- Human Parainfluenza virus 1, HPIV-1 (most common cause of croup; also other upper and lower respiratory tract illnesses typical)
- Human Parainfluenza virus 2, HPIV-2 (causes croup and other upper and lower respiratory tract illnesses)
- Human Parainfluenza virus 3, HPIV-3 (associated with bronchiolitis and pneumonia)
- Human Parainfluenza virus 4, HPIV-4 (mild respiratory infections, often clinically silent)
- Respiratory syncytial virus, RSV (pneumonia, pneumovirus pneumonia)

Rhabdoviridae:

- Duvenhage (rabies-like encephalitis)
- Rabies virus (rabies, hydrophobia, lyssa)
- Vesicular stomatitis virus (flu-like illness in humans)
- Chandipura (encephalitis)

Filoviridae:

- Ebola Reston (Ebola hemorrhagic fever)
- Ebola Siena (Ebola hemorrhagic fever)
- Ebola Sudan (Ebola hemorrhagic fever)
- Ebola Zaire (Ebola hemorrhagic fever)
- Marburg virus (Marburg hemorrhagic fever)

Bornaviridae:

Bornavirus, alternately Borna disease virus (Borna disease in mammals, possible cause of mental illness, including bipolar disorder in humans)

Arenaviridae:

Lassa virus (Lassa fever, multisystem disease characterized by hyperpyrexia, coagulopathy, hemorrhaging, and necrosis of liver and spleen)
Lujo virus (viral hemorrhagic fever)
Lymphocytic choriomeningitis virus (lymphocytic choriomeningitis)
Chapare virus (hemorrhagic fever) [137]
Flexal virus (flu-like illness)
Guanarito virus (Venezuelan hemorrhagic fever)
Junin virus (Argentine hemorrhagic fever)
Machupo virus (Bolivian hemorrhagic fever)
Oliveros virus (severe hemorrhagic fever in South America) [138]
Sabia virus (Brazilian hemorrhagic fever)
Whitewater Arroyo virus (hemorrhagic fever)

Bunyaviridae:

Bunyamwera virus (headache systemic symptoms, and rash)
California encephalitis virus (encephalitis)
Hantavirus Belgrade, also known as Hantavirus Dobrava-Belgrade (hemorrhagic fever with renal syndrome)
Hantavirus Hantaan (Korean hemorrhagic fever, hemorrhagic fever with renal syndrome)
Hantavirus Prospect Hill (not associated with human disease)
Hantavirus Puumala (hemorrhagic fever with renal syndrome)
Hantavirus Seoul (hemorrhagic fever with renal syndrome)
Hantavirus Sin Nombre (formerly Muerto Canyon)
Nairovirus Bhanja (rare, flu-like illness ranging from mild to severe disease)
Nairovirus Crimean (Congo haemorrhagic fever, Crimean-Congo hemorrhagic fever)
Nairovirus Hazara (hemorrhagic fever)
Oropouche virus (Oroopouche fever, characterized by fever with systemic symptoms)
RVF virus (Rift Valley fever)
Pappataci fever virus (Pappataci fever, phlebotomus fever, sandfly fever, three-day fever)

Orthomyxoviridae:

Influenza, types A, B and C (viral influenza)
Tick-borne orthomyxoviridae Dhori (fever, encephalitis)
Tick-borne orthomyxoviridae Thogoto (respiratory disease)

Deltavirus:

Hepatitis delta virus (high mortality hepatitis when co-infected or superinfected with Hepatitis B virus)

Chapter 44

Group VI Viruses: Single-Stranded RNA Reverse Transcriptase Viruses with a DNA Intermediate in Life-Cycle

"The origin of retroviruses is lost in a prebiotic mist."

Patric Jern, Goran Sperber, Jonas Blomberg [121]

Group I, dsDNA (Chapter 39)
Group II, ssDNA (Chapter 40)
Group III, dsRNA (Chapter 41)
Group IV (+)ssRNA (Chapter 42)
Group V (−)ssRNA (Chapter 43)
Group VI, ssRNA-RT (Chapter 44)
  Retroviridae
    Deltaretrovirus
      *Human T-cell lymphotropic viruses
    Lentivirus
      *Human immunodeficiency virus
Group VII, dsDNA-RT (Chapter 45)
Prions (Chapter 46)

Single-stranded RNA viruses can be positive sense or negative sense. Positive sense RNA, for example eukaryotic mRNA and Group IV viruses, can be directly translated to produce protein. Negative sense RNA is complementary to mRNA. Negative sense RNA must be copy-converted to positive sense RNA or to DNA before it becomes biologically available for translation or replication, respectively.

The Group VI viruses (Chapter 44) are single-stranded negative sense RNA viruses that use an RNA-dependent DNA polymerase (so-called reverse transcriptase), packaged within the virus particle, to produce a complementary

J.J. Berman: Taxonomic Guide to Infectious Diseases. DOI: http://dx.doi.org/10.1016/B978-0-12-415895-5.00044-1

strand of DNA. The synthesized strand of DNA is subsequently used as a template to yield a double-stranded DNA molecule containing the genetic information from the viral genome. Group VI viruses can integrate this double-stranded DNA into the host genome. The Group VI viruses are referred to as retroviruses.

The Group V viruses (Chapter 43), like the Group VI viruses, are single-stranded negative sense RNA viruses. These viruses do not employ reverse transcriptase. Instead, they use an RNA-dependent RNA polymerase, packaged within the virus particle, to produce positive sense RNA, within the host cell. The positive sense RNA is subsequently used to synthesize proteins.

Group VI viruses share their genetic legacy with the human genome. About 8% of human genes are retroviral. Human DNA of retroviral origin is referred to as endogenous retrovirus, or as a retroviral provirus. Retroviruses in the external environment, capable of infecting eukaryotic host cells, are referred to as exogenous retrovirus (i.e., a retrovirus that is outside the gene).

Despite the legacy of retroviruses within the genome of eukaryotic cells, there are only a few exogenous retroviruses that cause disease in humans. The Group VI human pathogens are restricted to one class of retroviruses, Class Retroviridae, and to two genera within this class: Deltaretrovirus and Lentivirus.

Group VI, ssRNA-RT
- Retroviridae
  - Deltaretrovirus
    - *Human T-cell lymphotropic viruses
  - Lentivirus
    - *Human immunodeficiency virus

Genus Deltaretrovirus contains four human T-cell lymphotropic viruses: HTLV-1, HTLV-2, HTLV-3, and HTLV-4. Of these four viruses that infect humans, only the HTLV-1 virus has been associated with human disease. Infection with HTLV-1 greatly increases the risk of developing adult T-cell leukemia/lymphoma, with about one out of every 25 infected individuals eventually developing the disease. HTLV-1 has also been implicated as a cause of a human myelopathic condition, tropical spastic paraparesis. Although millions of individuals have been infected by HTLV-1, worldwide, fewer than 2% of infected individuals will develop an HTLV-1-associated myelopathic condition.

Genus Lentivirus contains the human immunodeficiency virus, which produces HIV infection and the syndrome of associated diseases known as AIDS.

Readers should not confuse HTLV-III, a virus discovered in 2005, and which is not known at this time to produce disease in infected humans, with an early name (long since abandoned) that was assigned to the HIV virus.

Infectious species:

Human T-cell lymphotropic viruses
HTLV-1 (some cases of T-cell leukemia and T-cell lymphoma in adults, HTLV-1-associated myelopathy/tropical spastic paraparesis or HAM/TSP)
Human immunodeficiency virus (HIV infection and AIDS)

Chapter 45

Group VII Viruses: Double Stranded DNA Reverse Transcriptase Viruses

"Whatever Nature has in store for mankind, unpleasant as it may be, men must accept, for ignorance is never better than knowledge."

Enrico Fermi

Group I, dsDNA (Chapter 39)
Group II, ssDNA (Chapter 40)
Group III, dsRNA (Chapter 41)
Group IV (+)ssRNA (Chapter 42)
Group V (−)ssRNA (Chapter 43)
Group VI, ssRNA-RT (Chapter 44)
Group VII, dsDNA-RT (Chapter 45)
  Hepadnaviridae
    Orthohepadnavirus
      *Hepatitis B
Prions (Chapter 46)

Group VII viruses are double-stranded DNA viruses that can integrate their DNA into the host genome, using a reverse transcriptase enzyme. The reverse transcriptase enzyme acts upon RNA, transcribed from viral DNA, as the template for genomic DNA. Hence, the Group VII viruses are an unusual type of retrovirus that do not belong to Class Retroviridae (Group VI, Chapter 44), because the genome is double-stranded DNA. Likewise, the Group VII viruses are not classed within the double-stranded DNA viruses (Group I, Chapter 39) because their replication requires the synthesis of an RNA intermediate.

Group VII, dsDNA-RT
  Hepadnaviridae
    Orthohepadnavirus
      *Hepatitis B

J.J. Berman: Taxonomic Guide to Infectious Diseases. DOI: http://dx.doi.org/10.1016/B978-0-12-415895-5.00045-3

Group VII has one class of viruses that is pathogenic in humans: Class Hepadnaviridae. Hepadnaviruses (short for HEPAtic DNA VIRUS) have a small, circular DNA genome. Class Hepadnaviridae contains one viral species that is pathogenic in humans: hepatitis B virus.

Hepatitis B infects more than 200 million people, worldwide, causing 2 million deaths each year. Deaths are due to acute or chronic hepatitis and to ensuing conditions, such as cirrhosis and hepatocellular carcinoma. Infection is spread from infected persons through contact with body fluids (e.g., sexual intercourse or through inoculation with contaminated needles or tattoo instruments or though the use of contaminated blood transfusion products) (see Glossary item, Blood contamination).

As a virus that can insert part of its genome into host DNA, you might expect that it would be oncogenic (able to produce cancers). You would be correct. Hepatocellular carcinoma occurs in about 10% of infected patients who develop chronic hepatitis. Hepatitis B is the only DNA-transforming virus that is not a Group I virus (Chapter 39). Sections of the viral genome, inserted into host DNA, persist in cells of the hepatocellular carcinomas that eventually develop.

In addition to causing hepatitis B, the hepatitis B virus is essential for the replication of hepatitis delta virus. Hepatitis delta virus (Hepatitis D virus) is an RNA virus of Group V (Chapter 43). The hepatitis delta virus is a defective virus that cannot replicate without the help of the hepatitis B virus, which produces the protein coat for hepatitis delta virus. A co-infection with hepatitis B and hepatitis D produces a more aggressive disease than that produced with hepatitis B alone [139].

Infectious species:

Hepatitis B (acute hepatitis, chronic hepatitis, cirrhosis, hepatocellular carcinoma)

Chapter 46

Prions

"Life is a concept."

Patrick Forterre [140]

Everyone has heard the aphorism: "One bad apple spoils the bunch". This trite adage seems to be the principle underlying the prion diseases. A prion is a misfolded protein that can somehow serve as a template for proteins of the same type to misfold, producing collections of non-functioning protein globs that accumulate, causing cells to degenerate. The cells of the body that are most vulnerable to prion-produced disease are the neurons of the brain. The reason for the particular sensitivity of neurons to prion disease relates to the limited ability of neurons to replicate (i.e., to replace damaged neurons with new neurons), re-connect (to replace damaged connections between a neuron and other cells), and to remove degenerated cells and debris.

The term "prion" was introduced in 1982, by Stanley Prusiner [141]. Prions are the only infectious agent that contain neither DNA or RNA. Though few scientists would consider prions to be organisms, living or otherwise, they are included here to ensure that readers are aware of these biological agents. Prions are not confined to mammals. They have been observed in fungi, where their accumulation does not seem to produce any deleterious effect, and may even be advantageous to the organism [142].

There are five known prion diseases of humans, and all of them produce encephalopathies characterized by decreasing cognitive ability and impaired motor coordination. At present, all of the prion disease are progressive and fatal.

Infectious species:

Kuru prion (Kuru)
CJD prion (Creutzfeldt–Jakob disease, CJD)
nvCJD prion, or vCJD prion, or Bovine Spongiform Encephalopathy prion (New variant Creutzfeldt–Jakob disease, vCJD, nvCJD)
GSS prion (Gerstmann-Sträussler-Scheinker syndrome, GSS)
FFI prion (Fatal familial insomnia, FFI)

J.J. Berman: Taxonomic Guide to Infectious Diseases. DOI: http://dx.doi.org/10.1016/B978-0-12-415895-5.00050-7

Appendix I

Additional Notes on Taxonomy

"The purpose of narrative is to present us with complexity and ambiguity."

Scott Turow

This section of the book was relegated to the Appendix, because it should only be read by stalwarts, who have studied Chapters 1 through 46, and who hunger for more.

A taxonomy is a theoretical construct. As such, it needs to be constantly tested to determine whether the defined properties of each class actually extend to all of the members of the class, and whether the hierarchies of the classes are true. As new information is acquired, taxonomies will need to change. Some of the recent changes in the taxonomy of living organisms have involved the highest classes in the hierarchy. The first division of the Eukaryotes was assigned to Class Bikonta and Class Unikonta (analogous to dicot and monocot division in Class Angiospermae) [69]. Class Protoctista has been dropped from the formal taxonomy. Class Rhizaria was introduced. Class Microsporidia was moved from Class Protoctista to Class Fungi. Class Chlorophyta was moved from Class Protoctista to Class Archaeplastida.

Troubles also arise on the species level. For example, *Enterobius vermicularis* is called pinworm in the USA and threadworm in the UK; while *Strongyloides stercoralis* is just the opposite (threadworm in the USA and pinworm in the UK). The only way to escape this trans-Atlantic confusion is to translate the common name of the organism back to its standard Latin binomial.

Here is a sampling of recent name changes, in species or genus:

Aggregatibacter actinomycetemcomitans, formerly *Actinobacillus actinomycetemcomitans*
Anaplasma phagocytophilum, formerly assigned two different species names: *Ehrlichia phagocytophilium* and *Ehrlichia equi* [26]
Aonchotheca philippinensis, formerly *Capillaria philippinensis*
Arcanobacterium haemolyticum, formerly *Corynebacterium haemolyticum*
Arcanobacterium pyogenes, formerly *Actinomyces pyogenes*
Bartonella quintana, formerly *Rochalimaea quintana*

Brachyspira pilosicoli, formerly *Serpulina pilosicoli*
Burkholderia mallei, formerly *Pseudomonas mallei*
Chlamydophila pneumoniae, formerly TWAR serovar
Cladophialophora bantiana, formerly *Xylohypha bantiana*
Cystoisospora belli, formerly *Isospora belli*
Elizabethkingia meningoseptica, formerly *Chryseobacterium meningosepticum*, formerly *Flavobacterium meningosepticum*
Encephalitozoon intestinalis, formerly *Septata intestinalis*
Fluoribacter bozemanae, formerly *Legionella bozemanae*
Gardnerella vaginalis, formerly *Corynebacterium vaginalis*, formerly *Haemophilus vaginalis*
Helicobacter pylori, formerly *Campylobacter pylori*
Klebsiella granulomatis, formerly *Calymmatobacterium granulomatis*, formerly *Donovania granulomatis*
Malassezia furfur, formerly *Pityrosporum ovale*
Micromonas micros, formerly *Peptostreptococcus micros* formerly *Parvimonas micros*
Mycolcadus corymbifera, formerly *Absidia corymbifera*
Neorickettsia sennetsu, formerly *Ehrlichia sennetsu*
Norovirus, formerly Norwalk virus
Pneumocystis jiroveci, formerly *Pneumocystis carinii*
Rotavirus, formerly gastroenteritis virus type B
Sarcocystis suihominis, formerly *Isospora hominis*
Stenotrophomonas maltophilia, formerly *Pseudomonas maltophilia*
Tropheryma whipplei, formerly *Tropheryma whippelii*
Volutella cinerescens, formerly *Psilonia cinerescens*

The organism now called *Rhodococcus equi* has had more than its share of name changes. These include: *Corynebacterium equi*, *Bacillus hoagii*, *Corynebacterium purulentus*, *Mycobacterium* equi, *Mycobacterium restrictum*, *Nocardia restricta*, and *Proactinomyces restrictus*.

Taxonomic instability impacts negatively on clinical practice. When the name of an organism changes, so must the name of the associated disease. Consider "*Allescheria boydii*." People infected with this fungal organism were said to suffer from the disease known as allescheriasis. When the organism's name was changed to *Petriellidium boydii*, the disease name was changed to petriellidosis. When the fungal name was changed, once more, to *Pseudallescheria boydii*, the disease name was changed to pseudallescheriosis [102]. Changes in the standard names of a fungus, appearing in the International Code of Botanical Nomenclature, should trigger concurrent changes in the standard nomenclatures of medicine, such as the World Health Organization's International Classification of Disease, and the National Library of Medicine's Medical Subject Headings, and a variety of specialized disease nomenclatures. Some of these nomenclatures update infrequently. When disease nomenclatures lag behind official taxonomy, errors in coding and reporting infectious diseases will ensue.

The confusion never ends. Here is a list of obfuscated terminology, guaranteed to mislead all but the most compulsive students.

Chapter 5, The Alpha Proteobacteria – Readers should be careful not to confuse Bartonella with the similar-sounding Bordetella (Beta Proteobacteria, Chapter 6).

Chapter 5, The Alpha Proteobacteria – Readers should be aware that brucellosis has been known by a great number of different names, including Mediterranean fever. Mediterranean fever, an arcane synonym for brucellosis, should not be confused with familial Mediterranean fever (a gene disorder characterized by fever and abdominal pain) or with Mediterranean anemia (a synonym for thalassemia).

Chapter 5, The Alpha Proteobacteria – Readers should be aware that Neorickettsia, despite its name, is not a type of Rickettsia (i.e., not a member of Class Rickettsiaceae). Neorickettsia is a member of Class Anaplasmataceae; hence, the disease it causes is an ehrlichiosis.

Chapter 5, The Alpha Proteobacteria – Readers should not be confused by the term "scrub typhus" for infection by *Orientia tsutsugamushi* (alternately named *Rickettsia tsutsutgamushi*). This disease is grouped as a "spotted fever," not a form of typhus.

Chapter 7, Gamma Proteobacteria – The term "dysentery" (from the Latin "dys" and Greek "dus," meaning bad, and the Greek "enterikos" meaning intestine) is often used to connote a specific disease, but dysentery is non-specific term that can be applied to any enteric disorder associated with severe or bloody diarrhea. Because the group of diseases known as "dysentery" are the most frequent cause of childhood morbidity and mortality, it is important to use the term correctly. In developed countries, the term "dysentery" most often refers to salmonellosis, while in less developed countries, "dysentery" usually refers to shigellosis (also called bacillary dysentery, another misnomer) [143]. Other bacterial causes for dysentery are: *Vibrio cholerae*, *Escherichia coli*, *Clostridium difficile*, *Salmonella*, *Campylobacter jejuni*, and *Yersinia enterocolitica*. Viruses that cause dysentery include Rotavirus and Norwalk virus. The term "amoebic dysentery" is usually reserved for gastroenteritis caused by *Entamoeba histolytica*.

Chapter 7, Gamma Proteobacteria – Granuloma venereum, caused by *Klebsiella granulomatis*, can be mistaken clinically with two other diseases that are characterized by genital ulcers: syphilis (*Treponema pallidum*, Class Spirochaetae, Chapter 9), and chancroid (*Haemophilus ducreyi*, Class Gamma Proteobacteria, Chapter 7)). Adding to the confusion, the syphilitic genital ulcer is known as a chancre and must be distinguished from chancroid. One last caveat. Granuloma venereum, caused by *Klebsiella granulomatis*, must not be confused with lymphogranuloma venereum, caused by *Chlamydia trachomatis* (Class Chlamydiae, Chapter 13).

Chapter 7, Gamma Proteobacteria – Readers should not confuse rhinoscleroma, caused by *Klebsiella rhinoscleromatis* (Class Gamma Proteobacteria, Chapter 7), with rhinosporidiosis, caused by *Rhinosporidium seeberi* (Choanozoa, Chapter 23).

Chapter 7, Gamma Proteobacteria – *Salmonella typhi* and *Salmonella paratyphi* cause typhoid and paratyphoid, respectively. Neither of these diseases should be confused with typhus, caused by *Rickettsia typhi* and *Rickettsia prowazekii*. Both diseases (typhoid and typhus) take their root from a Greek word meaning stupor, referring to the neurologic manifestations of the diseases.

Chapter 7, Gamma Proteobacteria – Species of Genus Shigella are found only in causes of shigellosis. Shigellosis is also called bacillary dysentery. Despite its name, readers should not assume that the cause of bacillary dysentery is a member of Class Bacilli (Chapter 12). The exclusive cause of bacillary dysentery is Shigella species, belonging to the Gamma Proteobacteria (Chapter 7).

Chapter 7, Gamma Proteobacteria – *Shigella boydii*, one of the causes of shigellosis, should not be confused with *Pseudallescheria boydii*, a fungus in Class Ascomycota (Chapter 46), one of many fungal organisms associated with the skin infection maduromycosis.

Chapter 7, Gamma Proteobacteria – Readers should not confuse *Plesiomonas shigelloides*, containing the species name "shigelloides," with the genus name "Shigella".

Chapter 7, Gamma Proteobacteria – *Haemophilus influenzae* causes pneumonitis, meningitis, conjunctivitis, otitis media, and bacteremia, in infants and young children. Its species name, infuenzae, was assigned when the bacteria was mistakenly thought to be the cause of influenza. Influenza, also known as the flu, is caused exclusively by the influenza virus, a Group V orthomyxovirus (Chapter 43).

Chapter 7, Gamma Proteobacteria – *Haemophilus parainfluenzae* causes some cases of endocarditis. Despite its name, *Haemophilus parainfluenzae* is not the cause of the disease known as parainfluenza. Parainfluenza is a type of croup (laryngotracheobronchitis), and about 75% of the cases of croup are caused by the parainfluenza virus, a Group V virus (Chapter 43).

Chapter 7, Gamma Proteobacteria – *Haemophilus ducreyi* is the cause of chancroid, a sexually transmitted disease. It must not be confused with *Klebsiella granulomatus*, in Class Enterobacteriaceae, the cause of granuloma inguinale.

Chapter 7, Gamma Proteobacteria – Readers should not confuse Genus Acinetobacter (Gamma Proteobacteria, Chapter 7), with Class Actinobacteria (Chapter 14).

Chapter 9, Spirochaetae – Rat-bite fever is caused by either Spirillum minus or *Streptobacillus moniliformus* (Class Fusobacteria, Chapter 10) [49]. Regardless of the causative organism, or the phylogenetic classes to which the organisms are assigned, the clinical symptoms are similar, as is the treatment.

Chapter 11, Mollicutes – Erysipelothrix contains one infectious species; *Erysipelothrix rhusiopathiae*, the cause of erysipeloid, a type of cellulitis (subcutaneous infection). Students should not confuse erysipeloid with the similar-sounding disease, erysipelas. Both erysipeloid and erysipelas produce cellulitis. Erysipelas is more common and, potentially, a more serious disease than erysipeloid. Erysipelas is caused by members of Genus Streptococcus (Class Bacilli, Chapter 12). Two additional similar-sounding skin conditions are erythrasma, characterized by brown scaly skin patches, caused by *Corynebacterium minutissimum* (Class Actinobacteria, Chapter 14), and erythema infectiosum, caused by Parvovirus B19 (Chapter 40). All four skin conditions are associated with reddened skin, and all four diseases take their root from the Greek, "erusi," meaning red.

Chapter 12, Class Bacilli plus Class Clostridia – The term "bacillary" is misleading. You might think that the adjective "bacillary" would be restricted to members of Class Bacilli, its subclasses, and Genus Bacillus. It seldom does. The word "bacillus" has its root in Latin, from "baculum", a rod or staff, so the name has been applied to the second term in the bionomial name of bacteria that do not belong to Class Bacilli.

An example of a species with "bacilla" in its name, that is not a member of Class Bacilli is *Bartonella bacilliformis* (the cause of verruga peruana). Examples of genera with "bacillus" in their name, that are not members of Class Bacilli, are: Actinobacillus and Streptobacillus. Genus Streptobacillus (Class Fusobacteria, Chapter 10), is a terminologic catastrophe, as it is not a sister genus to Streptococcus, and it is not a member of Class Bacilli.

The subclasses of Class Bacilli were assigned based on phylogeny, not on morphology. Therefore, there are members of Class Bacilli that are not rod-shaped (e.g., Genus Staphylococcus and Streptococcus).

Furthermore, there are diseases containing the term "bacillary" that are not caused by members of Class Bacilli. These include bacillary angiomatosis (caused by *Bartonella quintana* and *Bartonella henselae*), and bacillary dysentery (caused by four different Shigella species and by *Yersinia enterocolitica*).

Chapter 12, Class Bacilli plus Class Clostridia – *Listeria monocytogenes* is the organism that causes listeriosis. Listeriosis should not be confused with the similar-sounding disease, leptospirosis (Spirochaetae, Chapter 9).

Chapter 13, Chlamydiae – Readers should not confuse trachoma with inclusion conjunctivitis, as each disease is caused by distinct variants of the same species (*Chlamydia trachomatis*). Trachoma is contracted by exposure to eye secretions from people with trachoma. Inclusion conjunctivitis is caused by ocular exposure to secretions from the sexually transmitted infection.

Chapter 13, Chlamydiae – *Chlamydia trachomatis* may also cause lymphogranuloma venereum, a disease that usually presents as swollen lymph nodes in the groin. The lymph nodes often have draining abscesses. The disease is rare, with only a few hundred cases occurring in the United States each year. Lymphogranuloma venereum must not be confused with granuloma inguinale, also known as granuloma venereum, caused by the bacterium *Klebsiella granulomatis* (Chapter 7).

Chapter 14, Actinobacteria – Members of Class Actinobacteria tend to be filamentous, and this morphologic feature led to great confusion. In the past, these filamentous bacteria were mistaken for fungal hyphae, and many of the diseases caused by members of Class Actinobacteria are given fungal names (e.g., actinomycosis, mycetoma, maduromycosis).

Chapter 14, Actinobacteria – Readers should be aware of the highly confusing term, "diphtheroid," from the Greek diphththera, meaning leather membrane, and commonly applied to all the non-pathogenic species within Genus Corynebacterium. As non-pathogens, the diphtheroids do not cause diphtheria. Diphtheria is caused *Corynebacterium diphtheriae* (i.e., a non-diphtheroid).

Chapter 14, Actinobacteria – Readers should not be confused by the plethora of organisms with "brasiliensis" as the species of the binomen. These include *Nocardia brasiliensis*, *Leishmania brasiliensis* (alternately *Leishmania braziliensis*), *Paracoccidioides brasiliensis*, and *Borrelia brasiliensis*.

Chapter 14, Actinobacteria – Readers should not confuse the bacterial genus Tropheryma with the similar-sounding term Taphrinomycotina, the fungal genus that includes Pneumocystis (Class Ascomycota, Chapter 36).

Chapter 19, Apicomplexa – The diseases produced by any of the organisms belonging to Class Coccidia are known collectively as coccidiosis. This term is applied most often to coccidian infections in animals, excluding humans. Coccidiosis must not be confused with the similar-sounding coccidioidomycosis (Ascomycota, Chapter 36).

Chapter 21, Heterokontophyta – Oomycota, a "colorless" class of heterokonts, contains the organisms that produce late blight of potato (*Phytophthora infestans*), and sudden oak death (*Phytophthora ramorum*). Oomycota, despite its suffix (mycota, an alternative name for fungus), is not a member of Class Fungi.

Chapter 21, Heterokontophyta – It is important to avoid confusing blastocystis with two similar terms used elsewhere in this book: blastomycosis and blastocyst. Blastomycoses is a fungal disease (Ascomycota, Chapter 36), and blastocyst is the blastula (Overview of Kingdom Animalia, Chapter 25) of mammals. All three terms come from the root word blastos (Greek for bud or embryo). Cystos (as in Blastocystis and blastocyst) is the Greek root meaning sac.

Chapter 22, Amoebozoa – Do not confuse *Entamoeba coli* (abbreviation *E. coli*) with *Escherichia coli* (likewise abbreviated as *E. coli*). Both live in the colon, and both can be reported in stool specimens. Do not confuse Entamoeba (Class Amoebozoa, Chapter 22) with Dientamoeba (Class Metamonada, Chapter 16).

Chapter 22, Amoebozoa – It is commonly agreed that the term "amoebiasis," with no qualifiers in the name, refers exclusively to the intestinal infection by *Entamoeba histolytica*. Encephalitides caused by members of Class Amoebozoa (Acanthamoeba and Balamuthia) are named granulomatous amoebic encephalitis. Encephalitis caused by *Naegleria fowleri* (not an amoeba) is called primary amoebic meningoencephalitis, an accepted misnomer. Naegleria is a member of class Percolozoa (Chapter 18). A better name for the Naeglerian disease would be primary percolozoan meningoencephalitis.

Chapter 24, Archaeplastida – Kleptoplast should not be confused with the similar-sounding word, kinetoplast. A kleptoplast is a chloroplast that has been stolen by another organism. A kinetoplast, uniquely found in members of Class Kinetoplastida, is a clump of DNA composed of copies of the mitochondrial genome, tucked inside a mitochondrion.

Chapter 27, Nematoda – Astute readers will notice that the prefix "trich" appears often within this chapter on Class Nematoda: Trichostrongylus, Trichocephalida, Trichinellidae, Trichinella, and Trichuris. A wide assortment of organisms, diseases, and medical terms contain the root "trich" (pronounced trick) and produce similar-sounding terms (i.e., homonyms). If you want to avoid confusing one disease with another, it is best to "come to terms" with this "trichy" nomenclature. The suffix "trich" comes from the Greek "thrix," meaning hair. Various unrelated organisms with a hair-like appendage, are provided with the "trich" suffix. Likewise, medical conditions of the hair are provided the same suffix: trichosis is any pathologic condition of hair; trichilemmoma is a tumor of hair, trichobezoar is the medical term for a hairball, trichotillomania is compulsive hair pulling. Words that sound somewhat like "trich" include trachoma (caused by the bacteria, *Chlamydia trachomatis*) and trachea (the windpipe). In addition to Trichomonas, there are several unrelated "trich" organisms that cause disease in humans: Trichinella, Trichomonas, Trichomonad, and Trichophyton. Other "trich" diseases include: *Trichostrongylosis trichinosis*, trichuriasis, trichomoniasis, trichiasis (everted eyelashes that touching the cornea or conjunctiva, often a post-infectious condition).

Chapter 27, Nematoda – When toxocara migrate through viscera, the condition is called visceral larva migrans. When toxocara migrate through an eye, the condition is called ocular larva migrans. When toxocara migrate through the skin, the condition is NOT called cutaneous larva migrans: this term is reserved for cutaneous manifestations of *Ancylostoma brasiliense*.

Chapter 27, Nematoda – Readers should not confuse "toxocara" with the similar-sounding "toxoplasma" (Class Apicomplexa, Chapter 19), a problem aggravated when clinicians use the abbreviated form "toxo," referring to "toxoplasmosis."

Chapter 27, Nematoda – Readers should avoid confusion by the layterms for Enterobius infection. In the United States, *Enterobius vermicularis* is known as pinworm; in the UK, it is known as threadworm. To add to the confusion, in the USA, *Strongyloides stercoralis* is known as threadworm: in the UK, it is known as pinworm.

Chapter 29, Chelicerata – Most members of Class Chelicerata are non-infectious in humans. Only two genera of Class Chelicerata live in, or on, humans, and both genera belong to the subclass of arachnids named Class Acari, which includes mites and ticks. Readers should not confuse mites and ticks with insects. Insects are members of Class Hexapoda (Chapter 30).

Chapter 30, Hexapoda – Class Hemiptera are the so-called "true bugs." They are distinguished from other insects by the shape of their mouth parts, which are shaped as a proboscis and covered by a labial sheath. The mouth parts of Class Hemiptera are designed for sucking. Class Hemiptera includes cicadas and aphids. The triatome species that are vectors for *Trypanosoma cruzi* (Euglenozoa, Chapter 17) are members of Class Hemiptera.

Chapter 31, Crustacea – One of the more confusing terms associated with pentastomiasis is "porocephaliasis," named for a pentastome genus, Porocephalus. The genus "Porocephalus" and the infection "porocephaliasis" should not be confused with "porocephaly," a rare developmental disorder in which cysts or cavities are found in the brains of infants.

Chapter 36, Ascomycota – Readers should be alerted that the term "Candida" is a source of some taxonomic confusion: candida, in Latin, means white. Many organisms are white, and have taken "candida" as part of a binomial name. Though there is only one Genus Candida (the fungus), there are many species named candida, particularly in Kingdom Plantae. For example, there are three *M. candida* species: *Mammilloydia candida*, a cactus; *Miltonia candida*, an orchid; and *Masdevallia candida*, another orchid.

Chapter 36, Ascomycota – Readers should remember not to confuse Microsporum with Microspora, a genus in Class Microsporaceae, a Chlorophyte. It is also important not to confuse Microsporum with the fungus of Genus Microsporidium (Class Microsporidia, Chapter 37).

Chapter 36, Ascomycota – Readers should not assume that Lobo's disease is caused by a member of Class Lobosea, a subclass of amoebozoans that includes Genus Acanthaemoeba (Chapter 22). Lobo's disease is caused by *Lacazia loboi*, an ascomycote fungus.

Chapter 37, Microsporidia – It is important not to confuse microsporidiosis with cryptosporidiosis, an apicomplexan disease (Apicomplexa, Chapter 19), that also produces diarrhea in immune-compromised patients.

Chapter 40, Group II Viruses – Bocavirus should not be confused with Bocas virus, a type of Coronavirus (Group IV, Chapter 42).

Chapter 42, Group IV Viruses – Hepevirus should not be confused with the orthographically similar "herpesvirus." Also, readers should not confuse Class Hepeviridae (hepatitis E virus) with Class Hepacivirus, a subclass of Class Flaviviridae that contains the Hepatitis C virus. Neither of these Group IV subclasses should be confused with Class Hepadnaviridae (Group VII, Chapter 45).

Chapter 42, Group IV Viruses – Readers should not confuse Rubella virus with the measles virus, Rubeola. Rubeola virus is a paramyxovirus (Group V, Chapter 43), unrelated to Rubella virus.

Chapter 43, Group V Viruses – One member of Class Arenaviridae, Lassa virus, the cause of Lassa fever, should not be confused with Lyssa virus, a member of Class Rhabdoviridae and the cause of rabies.

Chapter 44, Group VI Viruses – Readers should not confuse HTLV-III, a virus discovered in 2005, and which is not known at this time to produce disease in infected humans, with the same name (long since abandoned) that was assigned to the HIV virus.

Appendix II

Number of Occurrences of Some Common Infectious Diseases

Chapter 5, The Alpha Proteobacteria – Between 1918 and 1922, epidemic, louse-borne, typhus (*Rickettsia prowazekii*) infected 30 million people, in Eastern Europe and Russia, accounting for about 3 million deaths [29].

Chapter 6, Beta Proteobacteria – Approximately 1.5 million new cases of gonorrhea occur annually in North America, where gonorrhea is the third most common sexually transmitted disease [32]. The second most common sexually transmitted disease is chlamydia (Class Chlamydiae, Chapter 13), with about 4 million new cases each year in North America [32]. The most common sexually transmitted disease is trichomoniasis (Class Metamonada, Chapter 16), with about 8 million new cases each year in North America [32].

Chapter 7, Gamma Proteobacteria – Plague, caused by *Yersinia pestis*, is not an extinct disease. Each year, several thousand cases of plague occur worldwide, resulting in several hundred deaths. Virtually all of the cases occur in Africa.

Chapter 7, Gamma Proteobacteria – Between 10 000 and 50 000 cases of Legionnaire's disease occur each year in the United States.

Chapter 13, Chlamydiae – Various estimates would suggest that worldwide, more than half a billion people are infected with one or another subtype of *Chlamydia trachomatis*.

Chapter 13, Chlamydiae – According to the World Health Organization, there are about 37 million blind persons, worldwide. Trachoma, caused by *Chlamydia trachomatis*, is the number one infectious cause of blindness and accounts for about 4% of these cases. The second most common infectious cause of blindness worldwide is *Onchocerca volvulus* (Class Nematoda, Chapter 27), accounting for about 1% of cases [28].

Chapter 14, Actinobacteria – About 2 billion people (of the world's 7 billion population) have been infected with *Mycobacterium tuberculosis*.

Chapter 14, Actinobacteria – Leprosy, also known as Hansen's disease, is caused by *Mycobacterium leprae* and *Mycobacterium lepromatosis.* In 2005, there were about 300 000 new cases reported, worldwide [60].

Chapter 17, Euglenozoa – The two infectious genera in Class Trypanistomida are Leishmania and Trypanosoma. Leishmania species cause leishmaniasis, a disease that infects about 12 million people worldwide. Each year, about 60 000 people die from the visceral form of the disease.

Chapter 17, Euglenozoa – *Trypanosoma brucei* is the cause of African trypanosomiasis (sleeping sickness). The reported numbers of cases are widely considered to be unreliable, but it has been estimated that infection with *Trypanosoma brucei* accounts for about 50 000 deaths each year.

Chapter 17, Euglenozoa – *Trypanosoma cruzi* is the cause of Chagas disease, also known as American trypanosomiasis. It affects about 8 million people [74].

Chapter 19, Apicomplexa – Genus Plasmodium is responsible for human and animal malaria. About 300–500 million people are infected with malaria, worldwide. About 2 million people die each year from malaria [4,76].

Chapter 19, Apicomplexa – The sister genus to Sarcocystis is Toxoplasma. About one third of the human population has been infected (i.e.,about 2.3 billion people) by the only species that produces human toxoplasmosis: *Toxoplasma gondii.*

Chapter 22, Amoebozoa – It is estimated that about 50 million people are infected by *Entamoeba histolytica*, with about 70 000 deaths per year, worldwide.

Chapter 26, Platyhelminthes (flatworms) – Fasciolopsiasis is caused by *Fasciolopsis buski*, a large (up to 7.5 cm. length) fluke that lives in the intestines of the primary host (pigs and humans). The number of humans infected is about 10 million.

Chapter 26, Platyhelminthes (flatworms) – *Clonorchis sinensis*, the Chinese liver fluke (also known as the Oriental liver fluke) infects about 30 million people.

Chapter 26, Platyhelminthes (flatworms) – *Paragonimus westermani*, along with dozens of less frequent species within Genus Paragonimus causes the condition known as paragonomiasis. About 22 million people are infected worldwide, with most cases occurring in Southeast Asia, Africa, and South America.

Chapter 26, Platyhelminthes (flatworms) – About 200 million people are infected by schistosomes (i.e., have some form of schistosomiasis).

Chapter 27, Nematoda – *Ascaris lumbricoides*, the cause of ascariasis, infects about 1.5 billion people worldwide, making it the most common helminth (worm) infection of humans [90].

Chapter 27, Nematoda – About 150 million people are infected by the filarial nematodes (genera Brugia, Loa, Onchocerca, Mansonella, and

Wuchereria) [91]. *Wuchereria bancroft* and *Brugia malayi*, together, infect about 120 million individuals [91]. Most cases occur in Africa and Asia.

Chapter 27, Nematoda – Class Anyclostomadea contains the two species responsible for nearly all cases of hookworm disease in humans: *Ancylostoma duodenale* and *Necator americanus*. Hookworms infect about 600 million people.

Chapter 29, Chelicerata – Scabies is an exceedingly common, global disease, with about 300 million new cases occurring annually.

Chapter 29, Chelicerata – Demodex is a tiny mite that lives in facial skin. Demodex mites can be found in the majority of humans.

Chapter 39, Group I Viruses: double-stranded DNA – The BK polyomavirus rarely causes disease in infected patients, and the majority of humans carry the latent virus.

Chapter 39, Group I Viruses: double-stranded DNA – The JC polyomavirus persistently infects the majority of humans, but it is not associated with disease in otherwise healthy individuals.

Chapter 39, Group I Viruses: double-stranded DNA – Smallpox is reputed to have killed about 300 million people in the twentieth century, prior to the widespread availability of an effective vaccine. Smallpox has been referred to as the greatest killer in human history.

Chapter 41, Group III Viruses: double-stranded RNA – In 2004, rotavirus infections accounted for about a half million deaths in young children, from severe diarrhea [5].

Chapter 42, Group IV Viruses: single-stranded (+) sense RNA – Yellow fever virus seems to have originated in Africa and spread to other continents in the mid-seventeenth century. It was responsible for hundreds of thousands of deaths in North America alone. Today, there are about 200 000 cases of yellow fever, worldwide, with about 30 000 deaths [134]. Most infections occur in Africa.

Chapter 42, Group IV Viruses: single-stranded (+) sense RNA – More than 50 million dengue virus infections occur each year, worldwide. Most infections are asymptomatic or cause only mild disease. A minority of cases are severe.

Chapter 43, Group V Viruses: single-stranded (–) sense RNA – In 2001, measles virus accounted for about 745 000 deaths.

Chapter 43, Group V Viruses: single-stranded (–) sense RNA – Seasonal influenza kills between a quarter million and a half million people worldwide, each year. In the USA seasonal influenza accounts for about 40 000 deaths annually. The global 1917–1918 influenza pandemic caused somewhere between 50 million and 100 million deaths.

Chapter 45, Group VII Viruses: double-stranded DNA reverse transcriptase viruses – Hepatitis B infects more than 200 million people, worldwide, causing 2 million deaths each year.

Appendix III

Organisms Causing Infectious Disease in Humans

"The botanist is he who can affix similar names to similar vegetables, and different names to different ones, so as to be intelligible to every one."

Carolus Linnaeus

A few hundred organisms account for virtually all of the infectious diseases that occur in humans, and all of these organisms are listed in the text of the book. In addition, there are many "case report" pathogens that account for just a few known instances of human disease. Some of these infections are isolated within a small and relatively uninhabited geographic area. Some are veterinary diseases that do not ordinarily infect humans, but can cross over to produce self-limited human disease when the infective load is high (e.g., when a handler is exposed to a large number of organisms). Some are passed by eating exotic and undercooked food. Many of the ultra-rare infectious diseases in humans are well-known to veterinary pathologists, who routinely see and treat these infections in animals.

Some of these infections are opportunistic and occur in a highly select subset of vulnerable individuals (immunosuppressed, diabetic, renal-compromised, neonatal, elderly, malnourished, etc.). For the most part, diseases that arise in immune-compromised individuals are either commensals (typically non-pathogenic organisms that live in our bodies), or sub-clinical environmental pathogens (organisms encountered in the environment that do not cause overt disease in the majority of healthy individuals).

The Appendix contains about 1400 organisms that are pathogenic in humans, collected from many different literature sources [144,145]. Has every infectious organism been listed in the appendix? Not at all. New pathogenic agents are constantly arising for a variety of reasons: improved diagnostic methods, demonstrating specific organisms that can cause a single disease or lesion; known organisms, previously thought to be benign, that are found to cause disease in immune-deficient individuals; very rare organisms, located in an isolated geographic location, that had previously escaped attention; newly described subspecies of previously known infectious species; previously non-pathogenic organisms that acquire a virulence factor; viral pathogens for non-human animals that acquire a mutation that confers transmissibility to humans or between humans; nomenclature changes, wherein an old name is replaced with a new name. Although completeness is a worthy goal, you can expect some omissions in this Appendix.

For each genus listed in the Appendix, the taxonomic hierarchy is provided. The value of the hierarchy is that readers can instantly see the succession of parent classes for the genus. If you know something about the parent classes, you also know something about the lower genera; that is the basic premise of this book. Hierarchies were produced by a short computer program, previously published by the author [1,146]. The program utilizes the ancestral information contained in the publicly available taxonomy file used by the European Bioinformatics Institute (EBI). The EBI taxonomy file contains over 400 000 species, and each record in the file contains the name of the species, an identifier for the species, and an identifier for the parent (superclass) for the species.

Here is part of a sample record in the publicly available Taxonomy file:

ID: 50
PARENT ID: 49
RANK: genus
SCIENTIFIC NAME: Chondromyces

The sample record provides an ID number for the entry organism, Chondromyces, and for its parent class. Since every organism and class has a parent (i.e., every parent has a parent up to the root of the classification), a computer program can reconstruct the full parental lineage for one organism, or a list of 1400 organisms. The EBI taxonomy file is available for public download through anonymous ftp, from the EBI server: ftp://ftp.ebi.ac.uk/pub/databases/taxonomy/.

The computer output is listed below. Each record contains the name of a genus known to contain infectious species. This is followed by the list of species, for the genus, that are infectious, and the ancestral lineage for the genus. Students and clinicians can inspect the organisms in the Appendix and quickly learn the higher taxa to which they apply. Applying knowledge of higher taxa will allow readers to infer something about the biology of the lower species, and will help direct the student to the most appropriate resources in the printed literature or on the Web.

Abiotrophia
Species: *Abiotrophia defectiva*
Hierarchy: Abiotrophia: Aerococcaceae: Lactobacillales: Bacilli: Firmicutes

Acanthamoeba
Species: *Acanthamoeba rhysoides, Acanthamoeba polyphaga, Acanthamoeba palestinensis, Acanthamoeba hatchetti, Acanthamoeba culbertsoni, Acanthamoeba castellani, Acanthamoeba astronyxis*

Hierarchy: Acanthamoeba: Acanthamoebidae: Centramoebida: Variosea: Amoebozoa

Acanthocephalus

Species: *Acanthocephalus rauschi, Acanthocephalus bufonis* [147]

Hierarchy: Acanthocephalus: Acanthocephala (thorny-headed worms)

Achillurbainia

Species: *Achillurbainia recondita, Achillurbainia nouveli*

Hierarchy: Achillurbainia: Orchipedidae: Echinostomida: Trematoda: Platyhelminthes

Achromobacter

Species: *Achromobacter xylosoxidans, Achromobacter piechaudii*

Hierarchy: Achromobacter: Alcaligenaceae: Burkholderiales: Beta Proteobacteria: Proteobacteria

Acidaminococcus

Species: *Acidaminococcus fermentans*

Hierarchy: Acidaminococcus: Acidaminococcaceae: Selenomonadales: Negativicutes: Firmicutes

Acinetobacter

Species: *Acinetobacter radioresistens, Acinetobacter lwoffii, Acinetobacter junii, Acinetobacter johnsonii, Acinetobacter haemolyticus, Acinetobacter calcoaceticus, Acinetobacter baumannii*

Hierarchy: Acinetobacter: Moraxellaceae: Pseudomonadales: Gamma Proteobacteria: Proteobacteria

Acremonium

Species: *Acremonium strictum, Acremonium roseogriseum, Acremonium recifei, Acremonium potronii, Acremonium kiliense, Acremonium falciforme, Acremonium curvulum, Acremonium alabamense*

Hierarchy: Acremonium: Hypocreaceae: Hypocreales: Ascomycota: Fungi

Acrophialophora

Species: *Acrophialophora fusispora*

Hierarchy: Acrophialophora: Mitosporic ascomycota: Ascomycota

Actinobacillus

Species: *Actinobacillus ureae, Actinobacillus suis, Actinobacillus pleuropneumoniae, Actinobacillus lignieresii, Actinobacillus hominis, Actinobacillus equuli*

Hierarchy: Actinobacillus: Pasteurellaceae: Pasteurellales: Gamma Proteobacteria: Proteobacteria

Actinomadura

Species: *Actinomadura pelletieri, Actinomadura madurae*

Hierarchy: Actinomadura: Thermomonosporaceae: Streptosporangineae: Actinomycetales: Actinobacteridae: Actinobacteria (class): Actinobacteria

Actinomyces

Species: *Actinomyces turicensis, Actinomyces radingae, Actinomyces odontolyticus, Actinomyces neuii, Actinomyces naeslundii,*

Actinomyces meyeri, Actinomyces israelii, Actinomyces gerencseriae, Actinomyces georgiae
Hierarchy: Actinomyces: Actinomycetaceae: Actinomycineae: Actinomycetales: Actinobacteridae: Actinobacteria (class): Actinobacteria

Aerococcus
Species: *Aerococcus viridans*
Hierarchy: Aerococcus: Aerococcaceae: Lactobacillales: Bacilli: Firmicutes

Aeromonas
Species: *Aeromonas veronii, Aeromonas sobria, Aeromonas hydrophila, Aeromonas caviae*
Hierarchy: Aeromonas: Aeromonadaceae: Aeromonadales: Gamma Proteobacteria: Proteobacteria

Alaria
Species: *Alaria marcianae, Alaria americana*
Hierarchy: Alaria: Alariaceae: Laminariales: Phaeophyceae: Heterokontophyta: Chromalveolata

Alcaligenes
Species: *Alcaligenes odorans*
Hierarchy: Alcaligenes: Alcaligenaceae: Burkholderiales: Beta Proteobacteria: Proteobacteria

Alternaria
Species: *Alternaria tenuissima, Alternaria stemphyloides, Alternaria longipes, Alternaria infectoria, Alternaria dianthicola, Alternaria chlamydospora, Alternaria alternata*
Hierarchy: Alternaria tenuissima complex: Alternaria: Mitosporic pleosporaceae: Pleosporaceae: Pleosporineae: Pleosporales: Pleosporomycetidae: Dothideomycetes: Dothideomyceta: Leotiomyceta: Pezizomycotina: Saccharomyceta: Ascomycota

Amphimerus
Species: *Amphimerus pseudofelineus* (ultra-rare in humans)
Hierarchy: Amphimerus: Opisthorchiidae: Opisthorchiida: Trematoda: Platyhelminthes

Amycolatopsis
Species: *Amycolatopsis orientalis*
Hierarchy: Amycolatopsis: Pseudonocardiaceae: Pseudonocardineae: Actinomycetales: Actinobacteridae: Actinobacteria (class): Actinobacteria

Anatrichosoma
Infectious species: (monkey pathogen, ultra-rare in humans), *Anatrichosoma cutaneum*
Hierarchy: Anatrichosoma: Trichuridae: Trichocephalida: Enoplia: Enoplea: Nematoda (NCBI)

Ancylostoma

Species: *Ancylostoma malayanum, Ancylostoma duodenale, Ancylostoma ceylanicum, Ancylostoma caninum, Ancylostoma braziliense*

Hierarchy: Ancylostoma: Ancylostomatidea: Strongylida: Secernentea: Nematoda

Anisakis

Species: *Anisakis simplex, Anisakis physeteris*

Hierarchy: Anisakis: Anisakidae: Ascaridoidea: Ascaridida: Nematoda

Aonchotheca

Species: *Aonchotheca philippinensis*, formerly *Capillaria philippinensis* (capillariasis)

Hierarchy: Aonchotheca: Trichinellidae: Trichurida: Enoplia: Adenophorea: Nematoda

Aphanoascus

Species: *Aphanoascus fulvescens*

Hierarchy: Aphanoascus: Onygenaceae: Onygenales: Eurotiomycetidae: Eurotiomycetes: Leotiomyceta: Pezizomycotina: Saccharomyceta: Ascomycota

Apophysomyces

Species: *Apophysomyces elegans*

Hierarchy: Apophysomyces: Mucoraceae: Mucorales: Mucoromycotina: Zygomycota

Arcanobacterium

Species: *Arcanobacterium pyogenes, Arcanobacterium heamolyticum, Arcanobacterium bernardiae*

Hierarchy: Arcanobacterium: Actinomycetaceae: Actinomycineae: Actinomycetales: Actinobacteridae: Actinobacteria (class): Actinobacteria

Arcobacter

Species: *Arcobacter cryaerophilus, Arcobacter butzleri*

Hierarchy: Arcobacter: Campylobacteraceae: Campylobacterales: Epsilon Proteobacteria: Delta/epsilon subdivisions: Proteobacteria

Arthrinium

Species: *Arthrinium phaeospermum*

Hierarchy: Arthrinium: Mitosporic apiosporaceae: Apiosporaceae: Sordariomycetidae incertae sedis: Sordariomycetidae: Sordariomycetes: Sordariomyceta: Leotiomyceta: Pezizomycotina: Saccharomyceta: Ascomycota

Artyfechinostomum [148]

Species: *Artyfechinostomum mehrai*

Hierarchy: Artyfechinostomum: Echinostomatidae: Echinostomata: Echinostomida: Trematoda: Platyhelminthes

Ascaris

Species: Ascaris suum, Ascaris lumbricoides

Hierarchy: Ascaris: Ascarididae: Ascaridoidea: Ascaridida: Nematoda

Ascocotyle (heterophyiosis)

Species: *Ascocotyle* sp.

Hierarchy: Ascocotyle: Heterophyidae: Opisthorchiida: Trematoda: Platyhelminthes

Aspergillus

Species: *Aspergillus wentii, Aspergillus versicolor, Aspergillus terreus* group, *Aspergillus oryzae, Aspergillus niger, Aspergillus nidulans* group, *Aspergillus glaucus, Aspergillus fumigatus* group, *Aspergillus flavus* group, *Aspergillus flavipes, Aspergillus fisherianus, Aspergillus clavatus, Aspergillus candidus*

Hierarchy: Aspergillus: Mitosporic trichocomaceae: Trichocomaceae: Eurotiales: Eurotiomycetidae: Eurotiomycetes: Leotiomyceta: Pezizomycotina: Saccharomyceta: Ascomycota

Aureobasidium

Species: *Aureobasidium pullulans*

Hierarchy: Aureobasidium: Mitosporic dothioraceae: Dothioraceae: Dothideales: Dothideomycetidae: Dothideomycetes: Dothideomyceta: Leotiomyceta: Pezizomycotina: Saccharomyceta: Ascomycota

Austrobilharzia (dermatitis)

Species: *Austrobilharzia terrigalensis* (bird schistosome)

Hierarchy: Austrobilharzia: Schistosomatidae: Schistosomatoidea: Strigeidida: Digenea: Trematoda: Platyhelminthes

Babesia

Species: *Babesia microti, Babesia gibsoni, Babesia divergens, Babesia bovis*

Hierarchy: Babesia: Babesiidae: Piroplasmida: Aconoidasida: Apicomplexa

Bacillus

Species: *Bacillus thuringiensis, Bacillus subtilis, Bacillus sphaericus, Bacillus pumilus, Bacillus mycoides, Bacillus licheniformis, Bacillus coagulans, Bacillus circulans, Bacillus cereus, Bacillus anthracis*

Hierarchy: Bacillus cereus group: Bacillus: Bacillaceae: Bacillales: Bacilli: Firmicutes

Bacteroides

Species: *Bacteroides vulgatus, Bacteroides ureolyticus, Bacteroides uniformis, Bacteroides thetaiotaomicron, Bacteroides stercoris, Bacteroides splanchnicus, Bacteroides pectinophilus, Bacteroides ovatus, Bacteroides merdae, Bacteroides galacturonicus, Bacteroides fragilis, Bacteroides forsythus, Bacteroides eggerthii, Bacteroides distasonis, Bacteroides caccae*

Hierarchy: Bacteroides: Bacteroidaceae: Bacteroidales: Bacteroidia: Bacteroidetes

Balamuthia

Species: *Balamuthia mandrillaris*

Hierarchy: Balamuthia: Balamuthiidae: Centramoebida: Amoebozoa

Balantidium

Species: *Balantidium coli*

Hierarchy: Balantidium: Balantidiidae: Vestibuliferida: Trichostomatia: Litostomatea: Intramacronucleata: Ciliophora: Alveolata

Bartonella

Species: *Bartonella quintana, Bartonella henselae, Bartonella elizabethae, Bartonella bacilliformis*

Hierarchy: Bartonella: Bartonellaceae: Rhizobiales: Alpha Proteobacteria: Proteobacteria

Basidiobolus

Species: *Basidiobolus ranarum* (basidiobolomycosis, formerly entomophthoromycosis)

Hierarchy: Basidiobolus: Basidiobolaceae: Entomophthorales: Zygomycetes: Zygomycota

Baylisascaris

Species: *Baylisascaris procyonis*

Hierarchy: Baylisascaris: Ascarididae: Ascaridoidea: Ascaridida: Nematoda

Beauveria

Species: *Beauveria bassiana*

Hierarchy: Cordyceps bassiana: Cordyceps: Cordycipitaceae: Hypocreales: Hypocreomycetidae: Sordariomycetes: Sordariomyceta: Leotiomyceta: Pezizomycotina: Saccharomyceta: Ascomycota

Bergeyella

Species: *Bergeyella zoohelcum*

Hierarchy: Bergeyella: Flavobacteriaceae: Flavobacteriales: Flavobacteria: Bacteroidetes

Bertiella

Species: *Bertiella studeri, Bertiella mucronata*

Hierarchy: Bertiella: Anoplocephalidae: Cyclophyllidea: Eucestoda: Cestoda: Platyhelminthes

Bifidobacterium

Species: *Bifidobacterium dentium*

Hierarchy: Bifidobacterium: Bifidobacteriaceae: Bifidobacteriales: Actinobacteridae: Actinobacteria (class): Actinobacteria

Bilharziella

Species: *Bilharziella polonica*

Hierarchy: Bilharziella: Schistosomatidae: Schistosomatoidea: Strigeidida: Digenea: Trematoda: Platyhelminthes

Bilophila

Species: *Bilophila wadsworthia*

Hierarchy: Bilophila: Desulfovibrionaceae: Desulfovibrionales: Delta Proteobacteria: Delta/epsilon subdivisions: Proteobacteria

Bipolaris

Species: *Bipolaris spicifera, Bipolaris hawaiiensis, Bipolaris australiensis*

Hierarchy: Bipolaris: Mitosporic cochliobolus: Pleosporaceae: Pleosporineae: Pleosporales: Pleosporomycetidae: Dothideomycetes: Dothideomyceta: Leotiomyceta: Pezizomycotina: Saccharomyceta: Ascomycota

Blastocystis

Species: *Blastocystis hominis*

Hierarchy: Blastocystis: Stramenopiles

Blastomyces

Species: *Blastomyces dermatitidis*

Hierarchy: Blastomyces: Ajellomycetaceae: Onygenales: Eurotiomycetes: Pezizomycotina: Ascomycota

Blastoschizomyces

Species: *Blastoschizomyces capitatus*

Hierarchy: Blastoschizomyces: Saccharomycetaceae: Saccharomycetales: Ascomycetes: Ascomycota: Fungi

Bolbosoma

Species: *Bolbosoma* sp.

Hierarchy: Bolbosoma: Corynosomatinae: Polymorphidae: Palaeacanthocephala: Acanthocephala: Pseudocoelomata: Bilateria: Animalia (from [149])

Bordetella

Species: *Bordetella pertussis, Bordetella parapertussis, Bordetella bronchiseptica, Bordetella avium*

Hierarchy: Bordetella: Alcaligenaceae: Burkholderiales: Beta Proteobacteria: Proteobacteria

Borrelia

Species: *Borrelia venezuelensis, Borrelia turicatae, Borrelia recurrentis, Borrelia persica, Borrelia parkeri, Borrelia mazzottii, Borrelia latyschewii, Borrelia hispanica, Borrelia hermsii, Borrelia duttonii, Borrelia crocidurae, Borrelia caucasica, Borrelia burgdorferi, Borrelia brasiliensis*

Hierarchy: Borrelia: Spirochaetaceae: Spirochaetales: Spirochaetes

Botryomyces

Species: *Botryomyces caespitosus*

Hierarchy: Botryomyces: Mitosporic dothideomycetes: Dothideomycetes: Dothideomyceta: Leotiomyceta: Pezizomycotina: Saccharomyceta: Ascomycota

Brevibacillus

Species: *Brevibacillus brevis*

Hierarchy: Brevibacillus: Paenibacillaceae: Bacillales: Bacilli: Firmicutes

Brevundimonas

Species: *Brevundimonas vesicularis, Brevundimonas diminuta*

Hierarchy: Brevundimonas: Caulobacteraceae: Caulobacterales: Alpha Proteobacteria: Proteobacteria

Brucella

Species: *Brucella melitensis*

Hierarchy: Brucella: Brucellaceae: Rhizobiales: Alpha Proteobacteria: Proteobacteria

Brugia

Species: *Brugia timori, Brugia pahangi, Brugia malayi, Brugia guyanensis, Brugia beaveri*

Hierarchy: Brugia: Onchocercidae: Filarioidea: Spirurida: Nematoda

Bunostomum

Species: *Bunostomum phlebotomum*

Hierarchy: Bunostomum: Bunostominae: Ancylostomatidae: Ancylostomatoidea: Strongylida: Rhabditida: Nematoda

Burkholderia

Species: *Burkholderia pseudomallei, Burkholderia mallei, Burkholderia cepacia*

Hierarchy: Pseudomallei group: Burkholderia: Burkholderiaceae: Burkholderiales: Beta Proteobacteria: Proteobacteria

Campylobacter

Species: *Campylobacter upsaliensis, Campylobacter sputorum, Campylobacter rectus, Campylobacter lari, Campylobacter jejuni, Campylobacter hyointestinalis, Campylobacter gracilis, Campylobacter fetus, Campylobacter curvus, Campylobacter concisus, Campylobacter coli*

Hierarchy: Campylobacter: Campylobacteraceae: Campylobacterales: Epsilon Proteobacteria: Delta/epsilon subdivisions: Proteobacteria

Candida

Species: *Candida zeylanoides, Candida viswanathii, Candida tropicalis, Candida rugosa, Candida parapsilosis, Candida norvegensis, Candida lusitaniae, Candida lipolytica, Candida lambica, Candida krusei, Candida kefyr, Candida intermedia, Candida haemulonis, Candida guilliermondii, Candida glabrata, Candida famata, Candida catenulata, Candida albicans*

Hierarchy: Candida: Mitosporic saccharomycetales: Saccharomycetales: Saccharomycetes: Saccharomycotina: Saccharomyceta: Ascomycota

Capnocytophaga

Species: *Capnocytophaga sputigena, Capnocytophaga ochracea, Capnocytophaga gingivalis, Capnocytophaga cynodegmi, Capnocytophaga canimorsus*

Hierarchy: Capnocytophaga: Flavobacteriaceae: Flavobacteriales: Flavobacteria: Bacteroidetes

Cardiobacterium
Species: *Cardiobacterium hominis*
Hierarchy: Cardiobacterium: Cardiobacteriaceae: Cardiobacteriales: Gamma Proteobacteria: Proteobacteria
Cathaemasia
Species: *Cathaemasia cabrerai* [150]
Hierarchy: Cathaemasia: Cathaemasiidae: Echinostomatoidea: Echinostomata: Echinostomida: Digenea: Trematoda: Platyhelminthes
Cedecea
Species: *Cedecea neteri, Cedecea lapagei, Cedecea davisae*
Hierarchy: Cedecea: Enterobacteriaceae: Enterobacteriales: Gamma Proteobacteria: Proteobacteria
Cellulomonas
Species: *Cellulomonas turbata, Cellulomonas cellulans*
Hierarchy: Cellulomonas: Cellulomonadaceae: Micrococcineae: Actinomycetales: Actinobacteridae: Actinobacteria
Centipeda
Species: *Centipeda periodontii*
Hierarchy: Centipeda: Veillonellaceae: Selenomonadales: Negativicutes: Firmicutes
Centrocestus
Species: *Centrocestus formosanus, Centrocestus armatus*
Hierarchy: Centrocestus: Heterophyidae: Opisthorchiata: Opisthorchiida: Digenea: Trematoda: Platyhelminthes
Cephaliophora
Species: *Cephaliophora irregularis*
Hierarchy: Cephaliophora: Mitosporic ascomycota: Ascomycota
Cerinosterus
Species: *Cerinosterus cyanescens*
Hierarchy: Cerinosterus: mitosporic Basidiomycota: Basidiomycota: Dikarya: Fungi
Chaetomium
Species: *Chaetomium strumarium, Chaetomium perpulchrum, Chaetomium globosum, Chaetomium funicola, Chaetomium atrobtunneum*
Hierarchy: Chaetomium: Chaetomiaceae: Sordariales: Sordariomycetidae: Sordariomycetes: Pezizomycotina: Ascomycota: Dikarya: Fungi
Chlamydia
Species: *Chlamydia trachomatis*
Hierarchy: Chlamydia: Chlamydiaceae: Chlamydiales: Chlamydiae (class): Chlamydiae
Chlamydoabsidia
Species: *Chlamydoabsidia padenii*

Hierarchy: Chlamydoabsidia: Mucoraceae: Mucorales: Mucoromycotina: Zygomycota

Chlamydophila

Species: *Chlamydophila psittaci, Chlamydophila pneumoniae*

Hierarchy: Chlamydophila: Chlamydiaceae: Chlamydiales: Chlamydiae (class): Chlamydiae

Chromobacterium

Species: *Chromobacterium violaceum*

Hierarchy: Chromobacterium: Chromobacterium group: Neisseriaceae: Neisseriales: Beta Proteobacteria: Proteobacteria

Chryseobacterium

Species: *Chryseobacterium meningosepticum, Chryseobacterium balustinum*

Hierarchy: Chryseobacterium: Flavobacteriaceae: Flavobacteriales: Flavobacteria: Bacteroidetes: Bacteria

Citrobacter

Species: *Citrobacter youngae, Citrobacter werkmanii, Citrobacter sedlakii, Citrobacter rodentium, Citrobacter koseri, Citrobacter freundii, Citrobacter farmeri, Citrobacter braakii, Citrobacter amalonaticus*

Hierarchy: Citrobacter: Enterobacteriaceae: Enterobacteriales: Gamma Proteobacteria: Proteobacteria

Cladophialophora

Species: *Cladophialophora devriesii, Cladophialophora carrionii, Cladophialophora boppii, Cladophialophora bantiana, Cladophialophora arxii*

Hierarchy: Cladophialophora: Mitosporic herpotrichiellaceae: Herpotrichiellaceae: Chaetothyriales: Chaetothyriomycetidae: Eurotiomycetes: Leotiomyceta: Pezizomycotina: Saccharomyceta: Ascomycota

Cladorrhinum

Species: *Cladorrhinum bulbillosum*

Hierarchy: Cladorrhinum: Lasiosphaeriaceae: Sordariales: Sordariomycetidae: Sordariomycetes: Sordariomyceta: Leotiomyceta: Pezizomycotina: Saccharomyceta: Ascomycota

Cladosporium (common allergen producing respiratory reactions, and allergic reactions, including asthma, and a putative cause of skin and nail infections)

Species: *Cladosporium sphaerosphermum, Cladosporium oxysporum, Cladosporium elatum, Cladosporium cladosporioides*

Hierarchy: Cladosporium: mitosporic Davidiellaceae: Davidiellaceae: Capnodiales: Dothideomycetidae: Dothideomycetes: Pezizomycotina: Ascomycota: Dikarya: Fungi

Clinostomum

Species: *Clinostomum complanatum*

Hierarchy: Clinostomum: Clinostomidae: Clinostomoidea: Strigeidida: Digenea: Trematoda: Platyhelminthes

Clostridium

Species: *Clostridium tetani, Clostridium tertium, Clostridium sporogenes, Clostridium sordellii, Clostridium septicum, Clostridium ramosum, Clostridium perfringens, Clostridium novyi, Clostridium histolyticum, Clostridium fallax, Clostridium difficile, Clostridium chauvoei, Clostridium butyricum, Clostridium botulinum, Clostridium bifermentans, Clostridium baratii*

Hierarchy: Clostridium: Clostridiaceae: Clostridiales: Clostridia: Firmicutes

Coccidioides

Species: *Coccidioides immitis*

Hierarchy: Coccidioides: Mitosporic onygenales: Onygenales: Eurotiomycetidae: Eurotiomycetes: Leotiomyceta: Pezizomycotina: Saccharomyceta: Ascomycota

Cokeromyces

Species: *Cokeromyces recurvartus*

Hierarchy: Cokeromyces: Thamnidiaceae: Mucorales: Mucoromycotina: Zygomycota

Colletotrichum

Species: *Colletotrichum gloeosporioides, Colletotrichum dematium, Colletotrichum coccodes*

Hierarchy: Glomerella cingulata: Glomerella: Glomerellaceae: Glomerellales: Hypocreomycetidae: Sordariomycetes: Sordariomyceta: Leotiomyceta: Pezizomycotina: Saccharomyceta: Ascomycota

Collinsella

Species: *Collinsella aerofaciens*

Hierarchy: Collinsella: Coriobacteriaceae: Coriobacterineae: Coriobacteriales: Coriobacteridae: Actinobacteria (class): Actinobacteria

Comamonas

Species: *Comamonas testosteroni*

Hierarchy: Comamonas: Comamonadaceae: Burkholderiales: Beta Proteobacteria: Proteobacteria

Conidiobolus

Species: *Conidiobolus lamprauges, Conidiobolus incongruus, Conidiobolus coronatus*

Hierarchy: Conidiobolus: Ancylistaceae: Entomophthorales: Entomophthoromycotina: Zygomycota

Coniothyrium

Species: *Coniothyrium fuckelii*

Hierarchy: Coniothyrium: Mitosporic pleosporales: Pleosporales: Pleosporomycetidae: Dothideomycetes: Dothideomyceta: Leotiomyceta: Pezizomycotina: Saccharomyceta: Ascomycota

Contracaecum

Species: *Contracaecum osculatum*

Hierarchy: Contracaecum: Anisakidae: Ascaridoidea: Ascaridida: Nematoda

Corynebacterium

Species: *Corynebacterium xerosis, Corynebacterium urealyticum, Corynebacterium ulcerans, Corynebacterium striatum, Corynebacterium pseudotuberculosis, Corynebacterium pseudodiphthericum, Corynebacterium propinquum, Corynebacterium minutissimum, Corynebacterium macginleyi, Corynebacterium kutscheri, Corynebacterium jeikeium, Corynebacterium diphtheriae, Corynebacterium bovis, Corynebacterium argentoratense, Corynebacterium afermentans*

Hierarchy: Corynebacterium: Corynebacteriaceae: Corynebacterineae: Actinomycetales: Actinobacteridae: Actinobacteria (class): Actinobacteria

Corynosoma

Species: *Corynosoma strumosum*

Hierarchy: Corynosoma: Polymorphidae: Polymorphida: Palaeacanthocephala: Acanthocephala

Coxiella

Species: *Coxiella burnetii*

Hierarchy: Coxiella: Coxiellaceae: Legionellales: Gamma Proteobacteria: Proteobacteria

Cryptococcus

Species: *Cryptococcus neoformans*

Hierarchy: Filobasidiella/cryptococcus neoformans species complex: Filobasidiella: Tremellaceae: Tremellales: Tremellomycetes: Agaricomycotina: Basidiomycota

Cryptocotyle

Species: *Cryptocotyle lingua*

Hierarchy: Cryptocotyle: Heterophyidae: Opisthorchiata: Opisthorchiida: Digenea: Trematoda: Platyhelminthes

Cryptosporidium

Species: *Cryptosporidium parvum*

Hierarchy: Cryptosporidium: Cryptosporidiidae: Eimeriorina: Eucoccidiorida: Coccidia: Apicomplexa

Cunninghamella

Species: *Cunninghamella bertholletiae*

Hierarchy: Cunninghamella: Cunninghamellaceae: Mucorales: Mucoromycotina: Zygomycota

Curvularia

Species: *Curvularia verucculosa, Curvularia senegalensis, Curvularia pallescens, Curvularia lunata, Curvularia geniculata, Curvularia clavata, Curvularia brachyspora*

Hierarchy: Curvularia: mitosporic Cochliobolus: Pleosporaceae: Pleosporales: Pleosporomycetidae: Dothideomycetes: Pezizomycotina: Ascomycota: Dikarya: Fungi

Cyclodontostomum
Species: *Cyclodontostomum purvisi*
Hierarchy: Cyclodontostomum: Chabertiidae: Strongyloidea: Strongylida: Rhabditida: Nematoda
Cyclospora
Species: *Cyclospora cayetanensis*
Hierarchy: Cyclospora: Eimeriidae: Eimeriorina: Eucoccidiorida: Coccidia: Apicomplexa
Cylindrocarpon
Species: *Cylindrocarpon vaginae, Cylindrocarpon lichenicola, Cylindrocarpon cyanescens*
Hierarchy: Cylindrocarpon: mitosporic Neonectria: Nectriaceae: Hypocreales: Hypocreomycetidae: Sordariomycetes: Pezizomycotina: Ascomycota: Dikarya: Fungi
Delftia
Species: *Delftia acidovorans*
Hierarchy: Delftia: Comamonadaceae: Burkholderiales: Beta Proteobacteria: Proteobacteria
Dermatophilus
Species: *Dermatophilus congolensis*
Hierarchy: Dermatophilus: Dermatophilaceae: Micrococcineae: Actinomycetales: Actinobacteridae: Actinobacteria (class): Actinobacteria
Dichelobacter
Species: *Dichelobacter nodosus*
Hierarchy: Dichelobacter: Cardiobacteriaceae: Cardiobacteriales: Gamma Proteobacteria: Proteobacteria
Dichotomophthoropsis
Species: *Dichotomophthoropsis nymphaerum*
Hierarchy: Dichotomophthora: mitosporic Ascomycota: Ascomycota: Dikarya: Fungi
Dicrocoelium
Species: *Dicrocoelium hospes, Dicrocoelium dendriticum*
Hierarchy: Dicrocoelium: Dicrocoeliidae: Plagiorchioidea: Plagiorchiata: Plagiorchiida: Digenea: Trematoda: Platyhelminthes
Dientamoeba
Species: *Dientamoeba fragilis*
Hierarchy: Dientamoeba: Monocercomonadidae: Trichomonadida: Parabasalia: Trichozoa: Metamonada
Dioctophyme
Species: *Dioctophyme renale*
Hierarchy: Dioctophyme: Dioctophymatidae: Dioctophymatoidea: Enoplia: Enoplea: Nematoda
Dipetalonema
Species: *Dipetalonema reconditum, Dipetalonema arbuta*

Hierarchy: Dipetalonema: Onchocercidae: Filarioidea: Spirurida: Nematoda

Diphyllobothrium

Species: *Diphyllobothrium theileri, Diphyllobothrium stemmacephalum, Diphyllobothrium scoticum, Diphyllobothrium pacificum, Diphyllobothrium orcini, Diphyllobothrium nihonkaiense, Diphyllobothrium mansonoides, Diphyllobothrium latum, Diphyllobothrium lanceolatum, Diphyllobothrium klebanovskii, Diphyllobothrium houghtoni, Diphyllobothrium hians, Diphyllobothrium erinaceieuropaei, Diphyllobothrium elegans, Diphyllobothrium dendriticum, Diphyllobothrium dalliae, Diphyllobothrium cordatum, Diphyllobothrium cameroni*

Hierarchy: Diphyllobothrium: Diphyllobothriidae: Pseudophyllidea: Eucestoda: Cestoda: Platyhelminthes

Diplogonoporus

Species: *Diplogonoporus fukuokaensis, Diplogonoporus brauni, Diplogonoporus balaenopterae*

Hierarchy: Diplogonoporus: Diphyllobothriidae: Pseudophyllidea: Eucestoda: Cestoda: Platyhelminthes

Diplostomum

Species: *Diplostomum spathaceum*

Hierarchy: Diplostomum: Diplostomatidae: Diplostomoidea: Strigeidida: Digenea: Trematoda: Platyhelminthes

Dipylidium

Species: *Dipylidium caninum*

Hierarchy: Dipylidium: Dipylidiidae: Cyclophyllidea: Eucestoda: Cestoda: Platyhelminthes

Dirofilaria

Species: *Dirofilaria ursi, Dirofilaria tenuis, Dirofilaria subdermata, Dirofilaria striata, Dirofilaria repens, Dirofilaria immitis*

Hierarchy: Dirofilaria: Onchocercidae: Filarioidea: Spirurida: Nematoda

Dracunculus

Species: *Dracunculus medinensis, Dracunculus insignis*

Hierarchy: Dracunculus: Dracunculidae: Dracunculoidea: Spirurida: Nematoda

Drechslera

Species: *Drechslera biseptata*

Hierarchy: Drechslera: Mitosporic pleosporaceae: Pleosporaceae: Pleosporineae: Pleosporales: Pleosporomycetidae: Dothideomycetes: Dothideomyceta: Leotiomyceta: Pezizomycotina: Saccharomyceta: Ascomycota

Echinochasmus

Species: *Echinochasmus perfoliatus, Echinochasmus jiufoensis, Echinochasmus japonicus*

Hierarchy: Echinochasmus: Echinostomatidae: Echinostomida: Trematoda: Platyhelminthes

Echinococcus

Species: *Echinococcus vogeli, Echinococcus oligarthus, Echinococcus multilocularis, Echinococcus granulosus*

Hierarchy: Echinococcus: Taeniidae: Cyclophyllidea: Eucestoda: Cestoda: Platyhelminthes

Echinoparyphium

Species: *Echinoparyphium recurvatum*

Hierarchy: Echinoparyphium: Echinostomatidae: Echinostomatoidea: Echinostomata: Echinostomida: Digenea: Trematoda: Platyhelminthes

Echinostoma

Species: *Echinostoma revolutum, Echinostoma malayanum, Echinostoma macrorchis, Echinostoma jassyense, Echinostoma ilocanum, Echinostoma hortense, Echinostoma echinatum, Echinostoma cinetorchis*

Hierarchy: Echinostoma: Echinostomatidae: Echinostomatoidea: Echinostomata: Echinostomida: Digenea: Trematoda: Platyhelminthes

Edwardsiella

Species: *Edwardsiella tarda, Edwardsiella hoshinae*

Hierarchy: Edwardsiella: Enterobacteriaceae: Enterobacteriales: Gamma Proteobacteria: Proteobacteria

Eggerthella

Species: *Eggerthella lenta*

Hierarchy: Eggerthella: Coriobacteriaceae: Coriobacterineae: Coriobacteriales: Coriobacteridae: Actinobacteria (class): Actinobacteria

Ehrlichia

Species: *Ehrlichia sennetsu, Ehrlichia phagocytophila, Ehrlichia ewingii, Ehrlichia equi, Ehrlichia chaffeensis*

Hierarchy: Ehrlichia: Anaplasmataceae: Rickettsiales: Alpha Proteobacteria: Proteobacteria: Bacteria

Eikenella

Species: *Eikenella corrodens*

Hierarchy: Eikenella: Neisseriaceae: Neisseriales: Beta Proteobacteria: Proteobacteria

Emmonsia

Species: *Emmonsia parva, Emmonsia crescens*

Hierarchy: Emmonsia: Mitosporic onygenales: Onygenales: Eurotiomycetidae: Eurotiomycetes: Leotiomyceta: Pezizomycotina: Saccharomyceta: Ascomycota

Encephalitozoon

Species: *Encephalitozoon intestinalis, Encephalitozoon hellem, Encephalitozoon cuniculi*

Hierarchy: Encephalitozoon: Unikaryonidae: Apansporoblastina: Microsporidia

Engyodontium
Species: *Engyodontium album*
Hierarchy: Engyodontium: Mitosporic ascomycota: Ascomycota
Entamoeba
Species: *Entamoeba moshkovskii, Entamoeba histolytica, Entamoeba chattoni*
Hierarchy: Entamoeba: Entamoebidae: Archamoebae: Amoebozoa
Enterobacter
Species: *Enterobacter sakazakii, Enterobacter hormaechei, Enterobacter gergoviae, Enterobacter cloacae, Enterobacter cancerogenus, Enterobacter asburiae, Enterobacter amnigenus, Enterobacter aerogenes*
Hierarchy: Enterobacter: Enterobacteriaceae: Enterobacteriales: Gamma Proteobacteria: Proteobacteria: Bacteria
Enterobius
Species: *Enterobius vermicularis, Enterobius gregorii* [151]
Hierarchy: Enterobius: Oxyuridae: Oxyuroidea: Oxyurida: Nematoda
Enterococcus
Species: *Enterococcus raffinosus, Enterococcus mundtii, Enterococcus hirae, Enterococcus gallinarum, Enterococcus flavescens, Enterococcus faecium, Enterococcus faecalis, Enterococcus durans, Enterococcus casseliflavus, Enterococcus avium*
Hierarchy: Enterococcus: Enterococcaceae: Lactobacillales: Bacilli: Firmicutes
Enterocytozoon
Species: *Enterocytozoon bieneusi*
Hierarchy: Enterocytozoon: Enterocytozoonidae: Apansporoblastina: Microsporidia
Epidermophyton
Species: *Epidermophyton floccosum*
Hierarchy: Epidermophyton: Mitosporic arthrodermataceae: Arthrodermataceae: Onygenales: Eurotiomycetidae: Eurotiomycetes: Leotiomyceta: Pezizomycotina: Saccharomyceta: Ascomycota
Erysipelothrix
Species: *Erysipelothrix rhusiopathiae*
Hierarchy: Erysipelothrix: Erysipelotrichaceae: Erysipelotrichales: Erysipelotrichi: Firmicutes
Escherichia
Species: *Escherichia coli*
Hierarchy: Escherichia: Enterobacteriaceae: Enterobacteriales: Gamma Proteobacteria: Proteobacteria
Eubacterium
Species: *Eubacterium timidum, Eubacterium tenue, Eubacterium sulci, Eubacterium saphenum, Eubacterium saburreum, Eubacterium rectale,*

Eubacterium nodatum, Eubacterium multiforme, Eubacterium moniliforme, Eubacterium limosum, Eubacterium cylindroides, Eubacterium contortum, Eubacterium combesii, Eubacterium brachy
Hierarchy: Eubacterium: Eubacteriaceae: Clostridiales: Clostridia: Firmicutes: Bacteria

Eurytrema
Species: *Eurytrema pancreaticum*
Hierarchy: Eurytrema: Dicrocoeliidae: Plagiorchioidea: Plagiorchiata: Plagiorchiida: Digenea: Trematoda: Platyhelminthes

Eustrongylides
Species: *Eustrongylides* sp.
Hierarchy: Eustrongylides: Dioctophymatidae: Dioctophymatoidea: Enoplia: Enoplea: Nematoda

Ewingella
Species: *Ewingella americana*
Hierarchy: Ewingella: Enterobacteriaceae: Enterobacteriales: Gamma Proteobacteria: Proteobacteria

Exophiala
Species: *Exophiala spinifera, Exophiala salmonis, Exophiala psychrophila, Exophiala pisciphila, Exophiala moniliae, Exophiala jeanselmei, Exophiala dermatitidis*
Hierarchy: Exophiala: Mitosporic herpotrichiellaceae: Herpotrichiellaceae: Chaetothyriales: Chaetothyriomycetidae: Eurotiomycetes: Leotiomyceta: Pezizomycotina: Saccharomyceta: Ascomycota

Exserohilum
Species: *Exserohilum macginnisii, Exserohilum longirostratum*
Hierarchy: Exserohilum: mitosporic Pleosporaceae: Pleosporaceae: Pleosporales: Pleosporomycetidae: Dothideomycetes: Pezizomycotina: Ascomycota

Fasciola
Species: *Fasciola indica, Fasciola hepatica, Fasciola gigantica*
Hierarchy: Fasciola: Fasciolidae: Echinostomata: Echinostomida: Trematoda: Platyhelminthes

Fasciolopsis
Species: *Fasciolopsis buski*
Hierarchy: Fasciolopsis: Fasciolidae: Echinostomatoidea: Echinostomata: Echinostomida: Digenea: Trematoda: Platyhelminthes

Fibricola [152]
Species: *Fibricola seoulensis*
Hierarchy: Fibricola: Diplostomidae: Holostomidae: Trematoda: Platyhelminthes

Fibrobacter [153]
Species: *Fibrobacter intestinalis*

Hierarchy: Fibrobacter: Fibrobacteraceae: Fibrobacterales: Fibrobacteres: Fibrobacter/Acidobacteria

Filifactor

Species: *Filifactor alocis*

Hierarchy: Filifactor: Peptostreptococcaceae: Clostridiales: Clostridia: Firmicutes

Finegoldia

Species: *Finegoldia magna*

Hierarchy: Finegoldia: Clostridiales family xi. incertae sedis: Clostridiales incertae sedis: Clostridiales: Clostridia: Firmicutes

Fluoribacter

Species: *Fluoribacter gormanii, Fluoribacter dumoffii, Fluoribacter bozemanae*

Hierarchy: Fluoribacter: Legionellaceae: Legionellales: Gamma Proteobacteria: Proteobacteria

Francisella

Species: *Francisella tularensis*

Hierarchy: Francisella: Francisellaceae: Thiotrichales: Gamma Proteobacteria: Proteobacteria

Fusarium

Species: *Fusarium verticillioides, Fusarium ventricosum, Fusarium subglutinans, Fusarium solani, Fusarium sacchari, Fusarium proliferatum, Fusarium pallidoroseum, Fusarium oxysporum, Fusarium nivale, Fusarium napiforme, Fusarium moniliforme, Fusarium incarnatum, Fusarium dimerum, Fusarium chlamydosporum, Fusarium aquaeductuum*

Hierarchy: Fusarium: mitosporic Hypocreales: Hypocreales: Hypocreomycetidae: Sordariomycetes: Pezizomycotina: Ascomycota

Fusobacterium

Species: *Fusobacterium varium, Fusobacterium ulcerans, Fusobacterium periodonticum, Fusobacterium nucleatum, Fusobacterium necrophorum, Fusobacterium mortiferum*

Hierarchy: Fusobacterium: Fusobacteriaceae: Fusobacteriales: Fusobacteria (class): Fusobacteria

Gardnerella

Species: *Gardnerella vaginalis*

Hierarchy: Gardnerella: Bifidobacteriaceae: Bifidobacteriales: Actinobacteridae: Actinobacteria (class): Actinobacteria

Gastrodiscoides

Species: *Gastrodiscoides hominis*

Hierarchy: Gastrodiscoides: Paramphistomidae: Paramphistomata: Echinostomida: Digenea: Trematoda: Platyhelminthes

Gemella

Species: *Gemella morbillorum*

Hierarchy: Gemella: Bacillales family xi. incertae sedis: Bacillales incertae sedis: Bacillales: Bacilli: Firmicutes

Giardia

Species: *Giardia duodenalis*

Hierarchy: Giardia: Hexamitidae: Diplomonadida: Fornicata: Metamonada

Gigantobilharzia

Species: *Gigantobilharzia sturniae, Gigantobilharzia huttoni*

Hierarchy: Gigantobilharzia: Schistosomatidae: Schistosomatoidea: Strigeidida: Digenea: Trematoda: Platyhelminthes

Gnathostoma

Species: *Gnathostoma spinigerum, Gnathostoma nipponicum, Gnathostoma hispidum, Gnathostoma doloresi*

Hierarchy: Gnathostoma: Gnathostomatidae: Gnathostomatoidea: Spirurida: Nematoda

Gongylonema

Species: *Gongylonema pulchrum*

Hierarchy: Gongylonema: Gongylonematidae: Spiruroidea: Spirurida: Nematoda

Gordonia

Species: *Gordonia terrae, Gordonia sputi, Gordonia rubropertincta, Gordonia bronchialis, Gordonia amarae*

Hierarchy: Gordonia: Gordoniaceae: Corynebacterineae: Actinomycetales: Actinobacteridae: Actinobacteria

Granulicatella, formerly Abiotrophia (endocarditis, septic arthritis)

Species: *Granulicatella adjacens, Granulicatella elegans*

Hierarchy: Granulicatella: Carnobacteriaceae: Lactobacillales: Bacilli: Firmicutes: Bacteria

Gymnophalloides (intestinal fluke, endemic in Korea, oyster, *Crassostrea gigas*, as source [154])

Species: *Gymnophalloides seoi*

Hierarchy: Gymnophalloides: Gymnophallidae: Digenea: Trematoda: Platyhelminthes

Haemonchus

Species: *Haemonchus contortus*

Hierarchy: Haemonchus: Haemonchinae: Haemonchidae: Trichostrongyloidea: Strongylida: Rhabditida: Nematoda

Haemophilus

Species: *Haemophilus segnis, Haemophilus paraphrophilus, Haemophilus parainfluenzae, Haemophilus parahaemolyticus, Haemophilus influenzae, Haemophilus haemolyticus, Haemophilus ducreyi, Haemophilus aphrophilus, Haemophilus actinomycetemcomitans*

Hierarchy: Haemophilus: Pasteurellaceae: Pasteurellales: Gamma Proteobacteria: Proteobacteria

Hafnia

Species: *Hafnia alvei*

Hierarchy: Hafnia: Enterobacteriaceae: Enterobacteriales: Gamma Proteobacteria: Proteobacteria

Haplorchis

Species: *Haplorchis yokogawai, Haplorchis vanissima, Haplorchis taichui, Haplorchis pumilo*

Hierarchy: Haplorchis: Heterophyidae: Opisthorchiata: Opisthorchiida: Digenea: Trematoda: Platyhelminthes

Helicobacter

Species: *Helicobacter pylori, Helicobacter pullorum, Helicobacter heilmannii, Helicobacter fennelliae, Helicobacter cinaedi*

Hierarchy: Helicobacter: Helicobacteraceae: Campylobacterales: Epsilon Proteobacteria: Delta/epsilon subdivisions: Proteobacteria

Heterobilharzia

Species: *Heterobilharzia americana*

Hierarchy: Heterobilharzia: Schistosomatidae: Schistosomatoidea: Strigeidida: Digenea: Trematoda: Platyhelminthes

Heterophyes

Species: *Heterophyes nocens, Heterophyes heterophyes, Heterophyes dispar*

Hierarchy: Heterophyes: Heterophyidae: Opisthorchiata: Opisthorchiida: Digenea: Trematoda: Platyhelminthes

Heterophyopsis

Species: *Heterophyopsis continua*

Hierarchy: Heterophyopsis: Heterophyidae: Opisthorchiida: Trematoda: Platyhelminthes

Histoplasma

Species: *Histoplasma capsulatum*

Hierarchy: Histoplasma: Ajellomycetaceae: Onygenales: Eurotiomycetes: Pezizomycotina: Ascomycota

Hortaea

Species: *Hortaea werneckii*

Hierarchy: Hortaea: Mitosporic dothideales: Dothideales: Dothideomycetidae: Dothideomycetes: Dothideomyceta: Leotiomyceta: Pezizomycotina: Saccharomyceta: Ascomycota

Human T-Lymphotropic Virus

Species: Human T-Lymphotropic Virus 1, Human Immunodeficiency Virus 2, Human Immunodeficiency Virus 1

Hierarchy: Group VI virus

Hymenolepis

Species: *Hymenolepis nana, Hymenolepis diminuta*

Hierarchy: Hymenolepis: Hymenolepididae: Cyclophyllidea: Eucestoda: Cestoda: Platyhelminthes

Hypoderaeum

Species: *Hypoderaeum conoideum*

Hierarchy: Hypoderaeum: Echinostomatidae: Echinostomatoidea: Echinostomata: Echinostomida: Digenea: Trematoda: Platyhelminthes

Inermicapsifer

Species: *Inermicapsifer madagascariensis*

Hierarchy: Inermicapsifer: Anoplocephalidae: Cyclophyllidea: Eucestoda: Cestoda: Platyhelminthes

Isoparorchis

Species: *Isoparorchis hypselobagri*

Hierarchy: Isoparorchis: Isoparorchiidae: Azygiida: Digenea: Trematoda: Platyhelminthes

Isospora, synonymous with Cystoisospora

Species: *Isospora belli*

Hierarchy: Isospora: Eimeriidae: Eimeriorina: Eucoccidiorida: Coccidia: Apicomplexa: Alveolata: Eukaryota

Kingella

Species: *Kingella kingae, Kingella denitrificans*

Hierarchy: Kingella: Neisseriaceae: Neisseriales: Beta Proteobacteria: Proteobacteria

Klebsiella

Species: *Klebsiella pneumoniae, Klebsiella oxytoca, Klebsiella ornithinolytica, Klebsiella granulomatis*

Hierarchy: Klebsiella: Enterobacteriaceae: Enterobacteriales: Gamma Proteobacteria: Proteobacteria

Kluyvera

Species: *Kluyvera cryocrescens, Kluyvera ascorbata*

Hierarchy: Kluyvera: Enterobacteriaceae: Enterobacteriales: Gamma Proteobacteria: Proteobacteria

Lactobacillus

Species: *Lactobacillus* sp.

Hierarchy: Lactobacillus: Lactobacillaceae: Lactobacillales: Bacilli: Firmicutes: Bacteria

Lasiodiplodia

Species: *Lasiodiplodia theobromae*

Hierarchy: Botryosphaeria rhodina: Botryosphaeria: Botryosphaeriaceae: Botryosphaeriales: Dothideomycetes incertae sedis: Dothideomycetes: Dothideomyceta: Leotiomyceta: Pezizomycotina: Saccharomyceta: Ascomycota

Lecythophora

Species: *Lecythophora hoffmannii*

Hierarchy: Lecythophora: Mitosporic coniochaetaceae: Coniochaetaceae: Coniochaetales: Sordariomycetidae: Sordariomycetes: Sordariomyceta: Leotiomyceta: Pezizomycotina: Saccharomyceta: Ascomycota

Legionella
Species: *Legionella wadsworthii, Legionella tucsonensis, Legionella sainthelensi, Legionella rubrilucens, Legionella pneumophila, Legionella oakridgensis, Legionella longbeachae, Legionella lansingensis, Legionella jordanis, Legionella hackeliae, Legionella feeleii, Legionella cincinnatiensis, Legionella cherrii, Legionella birminghamensis, Legionella anisa*
Hierarchy: Legionella: Legionellaceae: Legionellales: Gamma Proteobacteria: Proteobacteria

Leifsonia
Species: *Leifsonia aquatica*
Hierarchy: Leifsonia: Microbacteriaceae: Micrococcineae: Actinomycetales: Actinobacteridae: Actinobacteria (class): Actinobacteria

Leishmania
Species: *Leishmania venezuelensis, Leishmania tropica, Leishmania shawi, Leishmania pifanoi, Leishmania peruviana, Leishmania panamensis, Leishmania naiffi, Leishmania mexicana, Leishmania major, Leishmania lainsoni, Leishmania infantum, Leishmania guyanensis, Leishmania donovani, Leishmania chagasi, Leishmania brasiliensis, Leishmania amazonensis, Leishmania aethiopica*
Hierarchy: Leishmania: Trypanosomatida: Kinetoplastida: Euglenozoa

Leptosphaeria
Species: *Leptosphaeria tompkinsii, Leptosphaeria senegalensis*
Hierarchy: Leptosphaeria: Leptosphaeriaceae: Pleosporineae: Pleosporales: Pleosporomycetidae: Dothideomycetes: Dothideomyceta: Leotiomyceta: Pezizomycotina: Saccharomyceta: Ascomycota

Leptospira
Species: *Leptospira weilii, Leptospira santarosai, Leptospira noguchii, Leptospira meyeri, Leptospira kirschneri, Leptospira interrogans, Leptospira inadai, Leptospira borgpetersenii*
Hierarchy: Leptospira: Leptospiraceae: Spirochaetales: Spirochaetes (class): Spirochaetes

Leptotrichia
Species: *Leptotrichia buccalis*
Hierarchy: Leptotrichia: Fusobacteriaceae: Fusobacteriales: Fusobacteria (class): Fusobacteria

Ligula
Species: *Ligula intestinalis*
Hierarchy: Ligula: Diphyllobothriidae: Pseudophyllidea: Eucestoda: Cestoda: Platyhelminthes

Listeria
Species: *Listeria welshimeri, Listeria seeligeri, Listeria monocytogenes, Listeria ivanovii*
Hierarchy: Listeria: Listeriaceae: Bacillales: Bacilli: Firmicutes

Loa
Species: *Loa loa*
Hierarchy: Loa: Onchocercidae: Filarioidea: Spirurida: Nematoda
Loboa (former name for Lacazia)
Species: *Loboa loboi*
Hierarchy: Lacazia (Loboa): Incertae sedis: Onygenales: Eurotiomycetes: Pezizomycotina: Ascomycota
Macracanthorhynchus
Species: *Macracanthorhynchus ingens, Macracanthorhynchus hirudinaceus*
Hierarchy: Macracanthorhynchus: Oligacanthorhynchidae: Oligacanthorhynchida: Archiacanthocephala: Acanthocephala
Madurella
Species: *Madurella mycetomati, Madurella grisea*
Hierarchy: Madurella: mitosporic Ascomycota: Ascomycota: Dikarya: Fungi
Malassezia
Species: *Malassezia sympodialis, Malassezia sloofiae, Malassezia restricta, Malassezia pachydermatis, Malassezia obtusa, Malassezia globosa, Malassezia furfur*
Hierarchy: Malassezia: Malasseziaceae: Malasseziales: Exobasidiomycetes: Ustilaginomycotina: Basidiomycota
Mammomonogamus
Species: *Mammomonogamus nasicola, Mammomonogamus laryngeus*
Hierarchy: Mammomonogamus: Syngamidae: Strongylida: Secernentea: Nematoda
Mannheimia
Species: *Mannheimia haemolytica*
Hierarchy: Mannheimia: Pasteurellaceae: Pasteurellales: Gamma Proteobacteria: Proteobacteria
Mansonella
Species: *Mansonella streptocerca, Mansonella semiclarum, Mansonella rodhaini, Mansonella perstans, Mansonella ozzardi*
Hierarchy: Mansonella: Onchocercidae: Spirurida: Secernentea: Nematoda
Marshallagia
Species: *Marshallagia marshalli*
Hierarchy: Marshallagia: Ostertagiinae: Haemonchidae: Trichostrongyloidea: Strongylida: Rhabditida: Nematoda
Mecistocirrus
Species: *Mecistocirrus digitatus*
Hierarchy: Mecistocirrus: Haemonchinae: Haemonchidae: Trichostrongyloidea: Strongylida: Rhabditida: Nematoda

Megamonas
Species: *Megamonas hypermegale*
Hierarchy: Megamonas: Veillonellaceae: Selenomonadales: Negativicutes: Firmicutes

Megasphaera
Species: *Megasphaera* sp.
Hierarchy: Megasphaera: Acidaminococcaceae: Clostridiales: Clostridia: Firmicutes: Bacteria

Meningonema
Species: *Meningonema peruzzii*
Hierarchy: Meningonema: Onchocercidae: Filarioidea: Spirurina: Spirurida: Secernentea: Nematoda (inferred hierarchy, clinical description in [92])

Mesocestoides
Species: *Mesocestoides variabilis, Mesocestoides lineatus*
Hierarchy: Mesocestoides: Mesocestoididae: Cyclophyllidea: Eucestoda: Cestoda: Platyhelminthes

Metagonimus
Species: *Metagonimus yokogawai, Metagonimus minutus*
Hierarchy: Metagonimus: Heterophyidae: Opisthorchiata: Opisthorchiida: Digenea: Trematoda: Platyhelminthes

Metastrongylus
Species: *Metastrongylus elongatus*
Hierarchy: Metastrongylus: Metastrongylidae: Metastrongyloidea: Strongylida: Rhabditida: Nematoda

Metorchis
Species: *Metorchis conjunctus, Metorchis albidus*
Hierarchy: Metorchis: Opisthorchiidae: Opisthorchioidea: Opisthorchiata: Opisthorchiida: Digenea: Trematoda: Platyhelminthes

Microfilaria (a taxonomically uncertified category for microfilariae that cannot be assigned to a known species)
Species: *Microfilaria bolivarensis*
Hierarchy: Microfilaria: Onchocercidae: Filarioidea: Spirurida: Nematoda

Micromonas
Infectious species: (periodontal disease); *Micromonas micros*, formerly *Peptostreptococcus micros,* alternately *Parvimonas micros*
Hierarchy: Micromonas: Clostridiaceae: Clostridiales: Clostridia: Firmicutes (inferred based on similarity to Peptostreptococcus)

Micronema
Infectious species: [155]; *Micronema deletrix*, alternately *Halicephalobus deletrix*
Hierarchy: Micronema: Panagrolaimidae: Rhabditida: Nematoda

Microsporum

Species: *Microsporum vanbreuseghemii, Microsporum racemosum, Microsporum cookei, Microsporum praecox, Microsporum persicolor, Microsporum nanum, Microsporum gypseum, Microsporum gallinae, Microsporum fulvum, Microsporum ferrugineum, Microsporum equinum, Microsporum canis, Microsporum audouinii*

Hierarchy: Microsporum: Mitosporic arthrodermataceae: Arthrodermataceae: Onygenales: Eurotiomycetidae: Eurotiomycetes: Leotiomyceta: Pezizomycotina: Saccharomyceta: Ascomycota

Moniezia

Species: *Moniezia expansa*

Hierarchy: Moniezia: Anoplocephalidae: Cyclophyllidea: Eucestoda: Cestoda: Platyhelminthes

Moniliella

Species: *Moniliella suavelolens*

Hierarchy: Moniliella: Incertae sedis: Ustilaginomycotina: Basidiomycota: Fungi

Moniliformis

Species: *Moniliformis moniliformis*

Hierarchy: Moniliformis: Moniliformidae: Moniliformida: Archiacanthocephala: Acanthocephala

Moraxella

Species: *Moraxella lincolnii, Moraxella osloensis, Moraxella nonliquefaciens, Moraxella liquefaciens, Moraxella lacunata, Moraxella bovis, Moraxella atlantae, Moraxella ovis, Moraxella cuniculi, Moraxella caviae, Moraxella catarrhalis*

Hierarchy: Moraxella: Moraxella: Moraxellaceae: Pseudomonadales: Gamma Proteobacteria: Proteobacteria

Morganella

Species: *Morganella morganii*

Hierarchy: Morganella: Enterobacteriaceae: Enterobacteriales: Gamma Proteobacteria: Proteobacteria

Mucor

Species: *Mucor ramosissimus, Mucor racemosus, Mucor indicus, Mucor hiemalis, Mucor circinelloides*

Hierarchy: Mucor: Mucoraceae: Mucorales: Mucoromycotina: Zygomycota

Multiceps (former name for Taenia)

Species: *Multiceps serialis, Multiceps multiceps, Multiceps longihamatus, Multiceps glomeratus, Multiceps brauni*

Hierarchy: Taenia: Taeniidae: Cyclophyllidea: Eucestoda: Cestoda: Platyhelminthes

Myceliophthora

Species: *Myceliophthora thermophila*

Hierarchy: Thielavia heterothallica: Thielavia: Chaetomiaceae: Sordariales: Sordariomycetidae: Sordariomycetes: Sordariomyceta: Leotiomyceta: Pezizomycotina: Saccharomyceta: Ascomycota

Mycobacterium

Species: *Mycobacterium xenopi, Mycobacterium ulcerans, Mycobacterium tuberculosis, Mycobacterium szulgai, Mycobacterium smegmatis, Mycobacterium simiae, Mycobacterium shimoidei, Mycobacterium senegalense, Mycobacterium scrofulaceum, Mycobacterium porcinum, Mycobacterium peregrinum, Mycobacterium mucogenicum, Mycobacterium marinum, Mycobacterium malmoense, Mycobacterium leprae, Mycobacterium kansasii, Mycobacterium haemophilum, Mycobacterium gordonae, Mycobacterium genavense, Mycobacterium fortuitum, Mycobacterium conspicuum, Mycobacterium chelonae, Mycobacterium celatum, Mycobacterium bovis, Mycobacterium avium, Mycobacterium asiaticum, Mycobacterium africanum, Mycobacterium abscessus*

Hierarchy: Mycobacterium: Mycobacteriaceae: Corynebacterineae: Actinomycetales: Actinobacteridae: Actinobacteria

Mycocentrospora

Species: *Mycocentrospora acerina*

Hierarchy: Mycocentrospora: Mitosporic mycosphaerellaceae: Mycosphaerellaceae: Capnodiales: Dothideomycetidae: Dothideomycetes: Dothideomyceta: Leotiomyceta: Pezizomycotina: Saccharomyceta: Ascomycota

Mycocladus

Species: *Mycolcadus corymbifera*, formerly *Absidia corymbifera*

Hierarchy: Mycocladus: Mycocladiaceae: Mucorales: Mucoromycotina: Zygomycota

Mycoleptodiscus

Species: *Mycoleptodiscus indicus*

Hierarchy: Mycoleptodiscus: Mitosporic magnaporthaceae: Magnaporthaceae: Magnaporthales: Sordariomycetidae: Sordariomycetes: Sordariomyceta: Leotiomyceta: Pezizomycotina: Saccharomyceta: Ascomycota

Mycoplasma

Species: *Mycoplasma salivarium, Mycoplasma pneumoniae, Mycoplasma hominis, Mycoplasma genitalium, Mycoplasma fermentans*

Hierarchy: Mycoplasma: Mycoplasmataceae: Mycoplasmatales: Mollicutes: Tenericutes

Myriodontium

Species: *Myriodontium keratinophilum*

Hierarchy: Myriodontium: Mitosporic onygenales: Onygenales: Eurotiomycetidae: Eurotiomycetes: Leotiomyceta: Pezizomycotina: Saccharomyceta: Ascomycota

Myroides
  Species: *Myroides odoratus*
  Hierarchy: Myroides: Flavobacteriaceae: Flavobacteriales: Flavobacteria: Bacteroidetes

Naegleria
  Species: *Naegleria fowleri*
  Hierarchy: Naegleria: Vahlkampfiidae: Schizopyrenida: Heterolobosea: Percolozoa

Nanophyetus
  Species: *Nanophyetus salmincola*
  Hierarchy: Nanophyetus: Nanophyetidae: Troglotremata: Plagiorchiida: Digenea: Trematoda: Platyhelminthes

Necator
  Species: *Necator americanus*
  Hierarchy: Necator: Bunostominae: Ancylostomatidae: Ancylostomatoidea: Strongylida: Rhabditida: Nematoda

Neisseria
  Species: *Neisseria weaveri, Neisseria subflava, Neisseria sicca, Neisseria perflava, Neisseria mucosa, Neisseria meningitidis, Neisseria lactamica, Neisseria gonorrhoeae, Neisseria flavescens, Neisseria flava, Neisseria elongata, Neisseria cinerea*
  Hierarchy: Neisseria: Neisseriaceae: Neisseriales: Beta Proteobacteria: Proteobacteria

Neocosmospora
  Species: *Neocosmospora vasinfecta*
  Hierarchy: Neocosmospora: Nectriaceae: Hypocreales: Hypocreomycetidae: Sordariomycetes: Sordariomyceta: Leotiomyceta: Pezizomycotina: Saccharomyceta: Ascomycota

Neodiplostomum
  Species: *Neodiplostomum* sp.
  Hierarchy: Neodiplostomum: Neodiplostomidae: Digenea: Trematoda: Platyhelminthes

Neotestudina
  Species: *Neotestudina rosatii*
  Hierarchy: Neotestudina: Testudinaceae: Pleosporales: Pleosporomycetidae: Dothideomycetes: Dothideomyceta: Leotiomyceta: Pezizomycotina: Saccharomyceta: Ascomycota

Nigrospora
  Species: *Nigrospora sphaerica*
  Hierarchy: Nigrospora: Mitosporic trichosphaeriales: Trichosphaeriales: Sordariomycetes incertae sedis: Sordariomycetes: Sordariomyceta: Leotiomyceta: Pezizomycotina: Saccharomyceta: Ascomycota

Nocardia

Species: *Nocardia transvalensis, Nocardia pseudobrasiliensis, Nocardia otitidiscaviarum, Nocardia nova, Nocardia farcinica, Nocardia caviae, Nocardia brasiliensis, Nocardia asteroides*

Hierarchy: Nocardia: Nocardiaceae: Corynebacterineae: Actinomycetales: Actinobacteridae: Actinobacteria (class): Actinobacteria

Nosema

Species: *Nosema ocularum, Nosema connori, Nosema ceylonensis, Nosema africanum*

Hierarchy: Nosema: Nosematidea: Microspora: Microsporidia: Fungi

Ochrobactrum

Species: *Ochrobactrum anthropi*

Hierarchy: Ochrobactrum: Brucellaceae: Rhizobiales: Alpha Proteobacteria: Proteobacteria

Ochroconis

Species: *Ochroconis gallopava*

Hierarchy: Ochroconis: Mitosporic ascomycota: Ascomycota

Oesophagostomum

Species: *Oesophagostomum stephanostomum, Oesophagostomum bifurcum, Oesophagostomum aculeatum*

Hierarchy: Oesophagostomum: Chabertiidae: Strongyloidea: Strongylida: Rhabditida: Nematoda

Oidiodendron

Species: *Oidiodendron cerealis*

Hierarchy: Oidiodendron: Mitosporic myxotrichaceae: Myxotrichaceae: Leotiomycetes incertae sedis: Leotiomycetes: Sordariomyceta: Leotiomyceta: Pezizomycotina: Saccharomyceta: Ascomycota

Oligella

Species: *Oligella urethralis, Oligella ureolytica*

Hierarchy: Oligella: Alcaligenaceae: Burkholderiales: Beta Proteobacteria: Proteobacteria

Onchocerca

Species: *Onchocerca volvulus*

Hierarchy: Onchocerca: Onchocercidae: Filarioidea: Spirurida: Nematoda

Onychocola

Species: *Onychocola canadensis*

Hierarchy: Onychocola: mitosporic Arachnomycetaceae: Arachnomycetaceae: Onygenales: Eurotiomycetidae: Eurotiomycetes: Pezizomycotina: Ascomycota: Dikarya: Fungi

Opisthorchis

Species: *Opisthorchis viverrini, Opisthorchis sinensis, Opisthorchis noverca, Opisthorchis guayaquilensis, Opisthorchis felineus*

Hierarchy: Opisthorchis: Opisthorchiidae: Opisthorchiata: Opisthorchiida: Digenea: Trematoda: Platyhelminthes

Orientia

Species: *Orientia tsutsugamushi* (scrub typhus)

Hierarchy: Orientia: Rickettsieae: Rickettsiaceae: Rickettsiales: Alpha Proteobacteria: Proteobacteria

Orientobilharzia

Species: *Orientobilharzia turkestanica*

Hierarchy: Orientobilharzia: Schistosomatidae: Schistosomatoidea: Strigeidida: Digenea: Trematoda: Platyhelminthes

Ornithobilharzia

Species: *Ornithobilharzia* sp.

Hierarchy: Ornithobilharzia: Schistosomatidae: Schistosomatoidea: Strigeidida: Digenea: Trematoda: Platyhelminthes

Ostertagia

Species: *Ostertagia ostertagi*

Hierarchy: Ostertagia: Ostertagiinae: Haemonchidae: Trichostrongyloidea: Strongylida: Rhabditida: Nematoda

Ovadendron

Species: *Ovadendron sulphureo-ochraceum, Ovadendron ochraceum* (fungal endophthalmitis)

Hierarchy: Ovadendron: Incertae sedis: Ascomycota

Paecilomyces (synonym for Isaria)

Species: *Paecilomyces viridis, Paecilomyces variotii, Paecilomyces marquandii, Paecilomyces lilacinus, Paecilomyces javanicus, Paecilomyces fumerosoreus, Paecilomyces farinosus*

Hierarchy: Isaria: mitosporic Cordycipitaceae: Cordycipitaceae: Hypocreales: Hypocreomycetidae: Sordariomycetes: Pezizomycotina: Ascomycota

Paenibacillus

Species: *Paenibacillus macerans, Paenibacillus alvei*

Hierarchy: Paenibacillus: Paenibacillaceae: Bacillales: Bacilli: Firmicutes

Pantoea

Species: *Pantoea agglomerans*

Hierarchy: Pantoea: Enterobacteriaceae: Enterobacteriales: Gamma Proteobacteria: Proteobacteria

Paracoccidioides

Species: *Paracoccidioides brasiliensis*

Hierarchy: Paracoccidioides: Mitosporic onygenales: Onygenales: Eurotiomycetidae: Eurotiomycetes: Leotiomyceta: Pezizomycotina: Saccharomyceta: Ascomycota

Paragonimus

Species: *Paragonimus westermani, Paragonimus uterobilateralis, Paragonimus skrjabini, Paragonimus siamensis, Paragonimus*

sadoensis, Paragonimus phillipinensis, Paragonimus ohirai, Paragonimus miyazakii, Paragonimus mexicanus, Paragonimus kellicotti, Paragonimus hueit'ungensis, Paragonimus heterotremus, Paragonimus caliensis, Paragonimus bankokensis, Paragonimus africanus
Hierarchy: Paragonimus: Paragonimidae: Troglotremata: Plagiorchiida: Digenea: Trematoda: Platyhelminthes

Parascaris
Species: *Parascaris equorum*
Hierarchy: Parascaris: Ascarididae: Ascaridoidea: Ascaridida: Nematoda

Parastrongylus (synonymous with Angiostrongylus)
Species: *Parastrongylus costaricensis, Parastrongylus cantonensis* (eosinophilic meningitis)
Hierarchy: Angiostrongylus: Angiostrongylidae: Metastrongyloidea: Strongylida: Rhabditida: Nematoda

Pasteurella
Species: *Pasteurella stomatis, Pasteurella pneumotropica, Pasteurella multocida, Pasteurella dagmatis, Pasteurella canis, Pasteurella caballi, Pasteurella aerogenes*
Hierarchy: Pasteurella: Pasteurellaceae: Pasteurellales: Gamma Proteobacteria: Proteobacteria

Pelodera
Species: *Pelodera strongyloides*
Hierarchy: Pelodera: Peloderinae: Rhabditidae: Rhabditoidea: Rhabditida: Nematoda

Penicillium
Species: *Penicillium purpurogenum, Penicillium marneffei, Penicillium expansum, Penicillium dupontii, Penicillium decumbens, Penicillium commune, Penicillium citrinum, Penicillium chrysogenum*
Hierarchy: Penicillium: Mitosporic trichocomaceae: Trichocomaceae: Eurotiales: Eurotiomycetidae: Eurotiomycetes: Leotiomyceta: Pezizomycotina: Saccharomyceta: Ascomycota

Pentatrichomonas
Species: *Pentatrichomonas hominis*
Hierarchy: Pentatrichomonas: Trichomonadidae: Trichomonadida: Parabasalia: Eukaryota

Peptococcus
Species: *Peptococcus niger*
Hierarchy: Peptococcus: Peptococcaceae: Clostridiales: Clostridia: Firmicutes

Peptostreptococcus
Species: *Peptostreptococcus vaginalis, Peptostreptococcus prevotii, Peptostreptococcus lactolyticus, Peptostreptococcus asaccharolyticus, Peptostreptococcus anaerobius*

Hierarchy: Peptostreptococcus: Peptostreptococcaceae: Clostridiales: Clostridia:
Firmicutes: Bacteria

Phaeoannellomyces (synonymous with Hortaea)
Species: *Phaeoannellomyces werneckii, Phaeoannellomyces elegans* (phaeohyphomycotic cyst)
Hierarchy: Hortaea: Dothioraceae: Dothideales: Dothideomycetidae: Dothideomycetes: Pezizomycotina: Ascomycota

Phaeosclera
Species: *Phaeosclera dematioides*
Hierarchy: Phaeosclera: Mitosporic ascomycota: Ascomycota

Phaeotrichoconis
Species: *Phaeotrichoconis crotalariae*
Hierarchy: Phaeotrichoconis: Mitosporic ascomycota: Ascomycota

Phaneropsolus [156]
Species: *Phaneropsolus* sp.
Hierarchy: Phaneropsolus: Lecithodendriidae: Plagiorchioidea: Plagiorchiata: Plagiorchiida: Digenea: Trematoda: Platyhelminthes

Phialemonium
Species: *Phialemonium obovatum, Phialemonium curvatum*
Hierarchy: Phialemonium: Mitosporic ascomycota: Ascomycota

Phialophora
Species: *Phialophora verrucosa, Phialophora richardsiae, Phialophora repens, Phialophora pedrosoi, Phialophora bubakii*
Hierarchy: Phialophora: Mitosporic magnaporthaceae: Magnaporthaceae: Magnaporthales: Sordariomycetidae: Sordariomycetes: Sordariomyceta: Leotiomyceta: Pezizomycotina: Saccharomyceta: Ascomycota

Philophthalmus
Species: *Philophthalmus lacrymosus*
Hierarchy: Philophthalmus: Philophthalmidae: Echinostomida: Trematoda: Platyhelminthes

Phocanema
Species: *Phocanema decipiens*
Hierarchy: Pseudoterranova: Pseudoterranova: Anisakidae: Ascaridoidea: Ascaridida: Nematoda

Phoma
Species: *Phoma oculo-hominis, Phoma minutella, Phoma hibernica, Phoma herbarum, Phoma glomerata, Phoma eupyrena, Phoma cruris-hominis, Phoma cava*
Hierarchy: Phoma: mitosporic Ascomycota: Ascomycota: Dikarya: Fungi

Photobacterium
Species: *Photobacterium damselae*

Hierarchy: Photobacterium: Vibrionaceae: Vibrionales: Gamma Proteobacteria: Proteobacteria

Physaloptera

Species: *Physaloptera transfuga, Physaloptera caucasica*

Hierarchy: Physaloptera: Physalopteridae: Physalopteroidea: Spirurida: Nematoda

Piedraia

Species: *Piedraia hortae*

Hierarchy: Piedraia: Piedraiaceae: Capnodiales: Dothideomycetidae: Dothideomycetes: Dothideomyceta: Leotiomyceta: Pezizomycotina: Saccharomyceta: Ascomycota

Plagiorchis

Species: *Plagiorchis philippinensis, Plagiorchis muris, Plagiorchis javensis, Plagiorchis harinasutai*

Hierarchy: Plagiorchis: Plagiorchiidae: Plagiorchioidea: Plagiorchiata: Plagiorchiida: Digenea: Trematoda

Plasmodium

Species: *Plasmodium vivax, Plasmodium simium, Plasmodium ovale, Plasmodium malariae, Plasmodium knowlesi, Plasmodium falciparum*

Hierarchy: Plasmodium (plasmodium): Plasmodium: Haemosporida: Aconoidasida: Apicomplexa

Plesiomonas

Species: *Plesiomonas shigelloides*

Hierarchy: Plesiomonas: Enterobacteriaceae: Enterobacteriales: Gamma Proteobacteria: Proteobacteria

Pleurophoma

Species: *Pleurophoma pleurospora*

Hierarchy: Pleurophoma: Mitosporic pleosporales: Pleosporales: Pleosporomycetidae: Dothideomycetes: Dothideomyceta: Leotiomyceta: Pezizomycotina: Saccharomyceta: Ascomycota

Pneumocystis

Species: *Pneumocystis carinii*

Hierarchy: Pneumocystis: Pneumocystidaceae: Pneumocystidales: Pneumocystidomycetes: Taphrinomycotina: Ascomycota

Porphyromonas

Species: *Porphyromonas macacae, Porphyromonas levii, Porphyromonas gingivalis, Porphyromonas endodontalis, Porphyromonas circumdentaria, Porphyromonas catoniae, Porphyromonas asaccharolytica*

Hierarchy: Porphyromonas: Porphyromonadaceae: Bacteroidales: Bacteroidia: Bacteroidetes

Prevotella

Species: *Prevotella zoogleoformans, Prevotella veroralis, Prevotella tannerae, Prevotella ruminicola, Prevotella oulora, Prevotella oris,*

Prevotella oralis, Prevotella nigrescens, Prevotella melaninogenica, Prevotella loescheii, Prevotella intermedia, Prevotella heparinolytica, Prevotella enoeca, Prevotella disiens, Prevotella denticola, Prevotella dentalis, Prevotella corporis, Prevotella buccalis, Prevotella buccae, Prevotella bivia
Hierarchy: Prevotella: Prevotellaceae: Bacteroidales: Bacteroidetes: Bacteroidetes/Chlorobi group: Bacteria

Propionibacterium
Species: *Propionibacterium propionicus, Propionibacterium granulosum, Propionibacterium avidum, Propionibacterium acnes*
Hierarchy: Propionibacterium: Propionibacteriaceae: Actinomycetales: Actinobacteridae: Actinobacteria

Proteus
Species: *Proteus vulgaris, Proteus penneri, Proteus mirabilis*
Hierarchy: Proteus: Enterobacteriaceae: Enterobacteriales: Gamma Proteobacteria: Proteobacteria

Prototheca
Species: *Prototheca zopfii, Prototheca wickerhamii*
Hierarchy: Prototheca: Chlorellaceae: Chlorellales: Trebouxiophyceae: Chlorophyta

Providencia
Species: *Providencia stuartii, Providencia rettgeri, Providencia alcalifaciens*
Hierarchy: Providencia: Enterobacteriaceae: Enterobacteriales: Gamma Proteobacteria: Proteobacteria

Pseudamphistomum
Species: *Pseudamphistomum truncatum, Pseudamphistomum aethiopicum*
Hierarchy: Pseudamphistomum: Opisthorchiidae: Opisthorchiata: Opisthorchiida: Digenea: Trematoda: Platyhelminthes

Pseudoallescheria
Species: *Pseudoallescheria boydii* (mucormycosis)
Hierarchy: Pseudallescheria: Microascaceae: Microascales: Sordariomycetes: Ascomycota

Pseudomicrodochium
Species: *Pseudomicrodochium suttonii*
Hierarchy: Pseudomicrodochium: Ascomycetes: Ascomycota

Pseudomonas
Species: *Pseudomonas stutzeri, Pseudomonas putida, Pseudomonas pseudoalcaligenes, Pseudomonas fluorescens, Pseudomonas alcaligenes, Pseudomonas aeruginosa*
Hierarchy: Pseudomonas stutzeri subgroup: Pseudomonas stutzeri group: Pseudomonas: Pseudomonadaceae: Pseudomonadales: Gamma Proteobacteria: Proteobacteria

Pseudonocardia

Species: *Pseudonocardia autotrophica*

Hierarchy: Pseudonocardia: Pseudonocardiaceae: Pseudonocardineae: Actinomycetales: Actinobacteridae: Actinobacteria (class): Actinobacteria

Pseudoramibacter

Species: *Pseudoramibacter alactolyticus*

Hierarchy: Pseudoramibacter: Eubacteriaceae: Clostridiales: Clostridia: Firmicutes

Psychrobacter

Species: *Psychrobacter phenylpyruvicus*

Hierarchy: Psychrobacter: Moraxellaceae: Pseudomonadales: Gamma Proteobacteria: Proteobacteria

Pygidiopsis

Species: *Pygidiopsis summa*

Hierarchy: Pygidiopsis: Heterophyidae: Opisthorchiata: Opisthorchiida: Digenea: Trematoda: Platyhelminthes

Pyrenochaeta

Species: *Pyrenochaeta unguis-hominis, Pyrenochaeta romeroi, Pyrenochaeta mackinnonii*

Hierarchy: Pyrenochaeta: Pleosporales incertae sedis: Pleosporales: Pleosporomycetidae: Dothideomycetes: Pezizomycotina: Ascomycota

Pythium

Species: *Pythium insidiosum* (pythiosis, wound infections)

Hierarchy: Pythium: Pythiaceae: Pythiales: Oomycetes: Heterokontophyta

Quaranfil

Species: Quaranfil Virus (febrile illness in children)

Hierarchy: Orthomyxoviridae, Group V

Rahnella

Species: *Rahnella aquatilis*

Hierarchy: Rahnella: Enterobacteriaceae: Enterobacteriales: Gamma Proteobacteria: Proteobacteria

Raillietina

Species: *Raillietina demerariensis, Raillietina celebensis*

Hierarchy: Raillietina: Davaineidae: Cyclophyllidea: Cestoda: Platyhelminthes

Ralstonia

Species: *Ralstonia pickettii*

Hierarchy: Ralstonia: Burkholderiaceae: Burkholderiales: Beta Proteobacteria: Proteobacteria

Rhinocladiella

Species: *Rhinocladiella schulzeri, Rhinocladiella obovoidea, Rhinocladiella compacta, Rhinocladiella aquaspersa, Rhinocladiella mackenziei* (chromoblastomycosis)

Hierarchy: Rhinocladiella: mitosporic Herpotrichiellaceae: Herpotrichiellaceae: Chaetothyriales: Chaetothyriomycetidae: Eurotiomycetes: Pezizomycotina: Ascomycota

Rhinosporidium

Species: *Rhinosporidium seeberi*

Hierarchy: Rhinosporidium: Dermocystida: Ichthyosporea: Choanozoa

Rhizomucor

Species: *Rhizomucor pusillus, Rhizomucor miehei*

Hierarchy: Rhizomucor: Mucoraceae: Mucorales: Mucoromycotina: Zygomycota

Rhizopus

Species: *Rhizopus stolonifer, Rhizopus oryzae, Rhizopus microsporus, Rhizopus azygosporus*

Hierarchy: Rhizopus: Mucoraceae: Mucorales: Mucoromycotina: Zygomycota

Rhodococcus

Species: *Rhodococcus rhodochrous, Rhodococcus rhodnii, Rhodococcus fascians, Rhodococcus erythropolis, Rhodococcus equi*

Hierarchy: Rhodococcus: Nocardiaceae: Corynebacterineae: Actinomycetales: Actinobacteridae: Actinobacteria (class): Actinobacteria

Rhodotorula

Species: *Rhodotorula rubra, Rhodotorula mucilaginosa, Rhodotorula minuta, Rhodotorula glutinis* (catheter infection)

Hierarchy: Rhodotorula: Sporidiobolaceae: Sporidiales: Urediniomycetes: Basidiomycota

Rickettsia

Species: *Rickettsia typhi, Rickettsia sibirica, Rickettsia rickettsii, Rickettsia prowazekii, Rickettsia massiliae, Rickettsia japonica, Rickettsia honei, Rickettsia felis, Rickettsia conorii, Rickettsia australis, Rickettsia akari, Rickettsia africae*

Hierarchy: Typhus group: Rickettsia: Rickettsieae: Rickettsiaceae: Rickettsiales: Alpha Proteobacteria: Proteobacteria

Rothia

Species: *Rothia dentocariosa*

Hierarchy: Rothia: Micrococcaceae: Micrococcineae: Actinomycetales: Actinobacteridae: Actinobacteria (class): Actinobacteria

Ruminococcus

Species: *Ruminococcus productus*

Hierarchy: Ruminococcus: Lachnospiraceae: Clostridiales: Clostridia: Firmicutes

Saccharomonospora

Species: *Saccharomonospora viridis*

Hierarchy: Saccharomonospora: Pseudonocardiaceae: Pseudonocardineae: Actinomycetales: Actinobacteridae: Actinobacteria (class): Actinobacteria

Saccharomyces

Species: *Saccharomyces cerevisiae*

Hierarchy: Saccharomyces: Saccharomycetaceae: Saccharomycetales: Saccharomycetes: Saccharomycotina: Saccharomyceta: Ascomycota

Saccharopolyspora

Species: *Saccharopolyspora rectivirgula*

Hierarchy: Saccharopolyspora: Pseudonocardiaceae: Pseudonocardineae: Actinomycetales: Actinobacteridae: Actinobacteria (class): Actinobacteria

Saksenaea

Species: *Saksenaea vasiformis*

Hierarchy: Saksenaea: Saksenaeaceae: Mucorales: Mucoromycotina: Zygomycota

Salmonella

Species: *Salmonella typhimurium, Salmonella typhi, Salmonella enteritidis, Salmonella choleraesuis, Salmonella bongori*

Hierarchy: Salmonella: Enterobacteriaceae: Enterobacteriales: Gamma Proteobacteria

Sarcinomyces

Species: *Sarcinomyces phaeomuriformis*

Hierarchy: Sarcinomyces: Mitosporic chaetothyriales: Chaetothyriales: Chaetothyriomycetidae: Eurotiomycetes: Leotiomyceta: Pezizomycotina: Saccharomyceta: Ascomycota

Sarcocystis

Species: *Sarcocystis suihominis, Sarcocystis lindermanni, Sarcocystis hominis*

Hierarchy: Sarcocystis: Sarcocystidae: Eimeriorina: Eucoccidiorida: Coccidia: Apicomplexa

Scedosporium

Species: *Scedosporium prolificans, Scedosporium apiospermum*

Hierarchy: Scedosporium: Mitosporic microascaceae: Microascaceae: Microascales: Hypocreomycetidae: Sordariomycetes: Sordariomyceta: Leotiomyceta: Pezizomycotina: Saccharomyceta: Ascomycota

Schistosoma

Species: *Schistosoma spindale, Schistosoma rodhaini, Schistosoma mekongi, Schistosoma mattheei, Schistosoma mansoni, Schistosoma malayensis, Schistosoma japonicum, Schistosoma intercalatum, Schistosoma haematobium, Schistosoma bovis*

Hierarchy: Schistosoma: Schistosomatidae: Schistosomatoidea: Strigeidida: Digenea: Trematoda: Platyhelminthes

Schistosomatium
Species: *Schistosomatium douthitti*
Hierarchy: Schistosomatium: Schistosomatidae: Schistosomatoidea: Strigeidida: Digenea: Trematoda: Platyhelminthes
Schizophyllum
Species: *Schizophyllum commune*
Hierarchy: Schizophyllum: Schizophyllaceae: Agaricales: Agaricomycetidae: Agaricomycetes: Agaricomycotina: Basidiomycota
Scolecobasidium
Species: *Scolecobasidium tshawytschae, Scolecobasidium humicola*
Hierarchy: Scolecobasidium: mitosporic Ascomycota: Ascomycota
Scopulariopsis (teleomorph Microascus)
Species: *Scopulariopsis fusca, Scopulariopsis cinereus (Microascus cinereus), Scopulariopsis flava, Scopulariopsis candida, Scopulariopsis brumptii, Scopulariopsis brevicaulis, Scopulariopsis asperula, Scopulariopsis acremonium*
Hierarchy: Scopulariopsis: Mitosporic microascales: Microascales: Hypocreomycetidae: Sordariomycetes: Sordariomyceta: Leotiomyceta: Pezizomycotina: Saccharomyceta: Ascomycota
Scytalidium
Species: *Scytalidium infestans, Scytalidium hyalinum*
Hierarchy: Scytalidium: incertae sedis: Helotiales: Leotiomycetes: Pezizomycotina: Ascomycota
Sebaldella
Species: *Sebaldella termitidis*
Hierarchy: Sebaldella: Fusobacteriaceae: Fusobacteriales: Fusobacteria (class): Fusobacteria
Selenomonas
Species: *Selenomonas noxia, Selenomonas infelix, Selenomonas flueggei, Selenomonas dianae, Selenomonas artemidis*
Hierarchy: Selenomonas: Veillonellaceae: Selenomonadales: Negativicutes: Firmicutes
Serratia
Species: *Serratia rubidaea, Serratia proteamaculans, Serratia plymuthica, Serratia odorifera, Serratia marcescens, Serratia ficaria*
Hierarchy: Serratia: Enterobacteriaceae: Enterobacteriales: Gamma Proteobacteria: Proteobacteria
Setaria
Species: *Setaria equina*
Hierarchy: Setaria: Setariidae: Filarioidea: Spirurida: Nematoda
Setosphaeria
Species: *Setosphaeria rostrata*

Hierarchy: Setosphaeria: Pleosporaceae: Pleosporineae: Pleosporales: Pleosporomycetidae: Dothideomycetes: Dothideomyceta: Leotiomyceta: Pezizomycotina: Saccharomyceta: Ascomycota

Shigella

Species: *Shigella sonnei, Shigella flexneri, Shigella dysenteriae, Shigella boydii*

Hierarchy: Shigella: Enterobacteriaceae: Enterobacteriales: Gamma Proteobacteria: Proteobacteria

Sphingomonas

Species: *Sphingomonas paucimobilis* (hospital-acquired infections)

Hierarchy: Sphingomonas: Sphingomonadaceae: Sphingomonadales: Alpha Proteobacteria: Proteobacteria

Spirillum

Species: *Spirillum minus* (rat bite fever, along with *Streptobacillus moniliformis* [49])

Hierarchy: Spirillum: Spirillaceae: Spirochaetales: Spirochaetes

Spirocerca

Species: *Spirocerca lupi*

Hierarchy: Spirocerca: Thelaziidae: Thelazioidea: Spirurida: Nematoda

Spirometra

Species: *Spirometra erinaceieuropaei*

Hierarchy: Spirometra: Diphyllobothriidae: Pseudophyllidea: Eucestoda: Cestoda: Platyhelminthes

Sporothrix

Species: *Sporothrix schenckii*

Hierarchy: Sporothrix: Mitosporic ophiostomataceae: Ophiostomataceae: Ophiostomatales: Sordariomycetidae: Sordariomycetes: Sordariomyceta: Leotiomyceta: Pezizomycotina: Saccharomyceta: Ascomycota

Staphylococcus

Species: *Staphylococcus warneri, Staphylococcus saprophyticus, Staphylococcus lugdunensis, Staphylococcus intermedius, Staphylococcus hyicus, Staphylococcus haemolyticus, Staphylococcus epidermidis, Staphylococcus aureus*

Hierarchy: Staphylococcus: Staphylococcaceae: Bacillales: Bacilli: Firmicutes

Stellantchasmus

Species: *Stellantchasmus falcatus*

Hierarchy: Stellantchasmus: Heterophyidae: Opisthorchiata: Opisthorchiida: Digenea: Trematoda: Platyhelminthes

Stenella

Species: *Stenella araguata*

Hierarchy: Stenella: Mitosporic mycosphaerellaceae: Mycosphaerellaceae: Capnodiales: Dothideomycetidae: Dothideomycetes: Dothideomyceta: Leotiomyceta: Pezizomycotina: Saccharomyceta: Ascomycota

Stenotrophomonas

Species: *Stenotrophomonas maltophilia*, formerly *Pseudomonas maltophilia* (pneumonia, urinary infection, bacteremia, from indwelling catheters and tubes)

Hierarchy: Stenotrophomonas maltophilia group: Stenotrophomonas: Xanthomonadaceae: Xanthomonadales: Gamma Proteobacteria: Proteobacteria

Streptobacillus

Species: *Streptobacillus moniliformis* (rat bite fever, along with *Spirillum minus* [49])

Hierarchy: Streptobacillus: Fusobacteriaceae: Fusobacteriales: Fusobacteria (class): Fusobacteria

Streptococcus

Species: *Streptococcus uberis, Streptococcus suis, Streptococcus sobrinus, Streptococcus sanguis, Streptococcus salivarius, Streptococcus pyogenes, Streptococcus pneumoniae, Streptococcus mutans, Streptococcus mitis, Streptococcus milleri, Streptococcus intermedius, Streptococcus gordonii, Streptococcus equi, Streptococcus criceti, Streptococcus constellatus, Streptococcus canis, Streptococcus bovis, Streptococcus anginosus, Streptococcus agalactiae, Streptococcus acidominimus*

Hierarchy: Streptococcus: Streptococcaceae: Lactobacillales: Bacilli: Firmicutes

Streptomyces

Species: *Streptomyces somaliensis, Streptomyces somaliensis* (mycetoma)

Hierarchy: Streptomyces: Streptomycetaceae: Streptomycineae: Actinomycetales: Actinobacteridae: Actinobacteria (class): Actinobacteria

Strongyloides

Species: *Strongyloides westeri, Strongyloides stercoralis, Strongyloides ransomi, Strongyloides papillosus, Strongyloides fuelleborni*

Hierarchy: Strongyloides: Strongyloididae: Panagrolaimoidea: Rhabditida: Nematoda

Sutterella

Species: *Sutterella wadsworthensis* (abscesses [157])

Hierarchy: Sutterella: Sutterellaceae: Burkholderiales: Beta Proteobacteria: Proteobacteria

Suttonella

Species: *Suttonella indologenes*

Hierarchy: Suttonella: Cardiobacteriaceae: Cardiobacteriales: Gamma Proteobacteria: Proteobacteria

Taenia

Species: *Taenia taeniaeformis, Taenia solium, Taenia saginata, Taenia crassiceps*

Hierarchy: Taenia: Taeniidae: Cyclophyllidea: Eucestoda: Cestoda: Platyhelminthes

Taeniolella

Species: *Taeniolella stilbospora, Taeniolella exilis*

Hierarchy: Taeniolella: Mitosporic botryosphaeriaceae: Botryosphaeriaceae: Botryosphaeriales: Dothideomycetes incertae sedis: Dothideomycetes: Dothideomyceta: Leotiomyceta: Pezizomycotina: Saccharomyceta: Ascomycota

Tatlockia

Species: *Tatlockia micdadei, Tatlockia maceachernii*

Hierarchy: Tatlockia: Legionellaceae: Legionellales: Gamma Proteobacteria: Proteobacteria

Tatumella

Species: *Tatumella ptyseos*

Hierarchy: Tatumella: Enterobacteriaceae: Enterobacteriales: Gamma Proteobacteria: Proteobacteria

Teladorsagia

Species: *Teladorsagia circumcincta*

Hierarchy: Teladorsagia: Ostertagiinae: Haemonchidae: Trichostrongyloidea: Strongylida: Rhabditida: Nematoda

Ternidens

Species: *Ternidens deminutus*

Hierarchy: Ternidens: Strongylidae: Strongyloidea: Strongylida: Rhabditida: Nematoda

Tetraploa

Species: *Tetraploa aristata*

Hierarchy: Tetraploa: Mitosporic tetraplosphaeriaceae: Tetraplosphaeriaceae: Pleosporales: Pleosporomycetidae: Dothideomycetes: Dothideomyceta: Leotiomyceta: Pezizomycotina: Saccharomyceta: Ascomycota

Thelazia (thelaziosis, eye infections, Diptera vector)

Species: *Thelazia rhodesii, Thelazia callipaeda, Thelazia californiensis*

Hierarchy: Thelazia: Thelaziidae: Spirurida: Secernentea: Nematoda

Thermomyces

Species: *Thermomyces lanuginosus*

Hierarchy: Thermomyces: Mitosporic ascomycota: Ascomycota

Torulopsis

Species: *Torulopsis magnoliae*, synonymous for *Candida magnoliae*

Hierarchy: Ascomycota: Saccharomycotina: Saccharomycetes: Saccharomycetales: Saccharomycetaceae: Candida (Torulopsis)

Toxocara
Species: *Toxocara cati, Toxocara canis*
Hierarchy: Toxocara: Toxocaridae: Ascaridoidea: Ascaridida: Nematoda
Toxoplasma
Species: *Toxoplasma gondii*
Hierarchy: Toxoplasma: Sarcocystidae: Eimeriorina: Eucoccidiorida: Coccidia: Apicomplexa
Trachipleistophora
Species: *Trachipleistophora hominis*
Hierarchy: Trachipleistophora: Pleistophoridae: Pansporoblastina: Microsporidia
Treponema
Species: *Treponema pallidum, Treponema carateum*
Hierarchy: Treponema: Spirochaetaceae: Spirochaetales: Spirochaetes (class): Spirochaetes
Trichinella
Species: *Trichinella spiralis, Trichinella pseudospiralis, Trichinella nelsoni, Trichinella nativa, Trichinella britovi*
Hierarchy: Trichinellidae: Trichocephalida: Enoplia: Enoplea: Nematoda
Trichobilharzia
Species: *Trichobilharzia stagnicolae, Trichobilharzia ocellata, Trichobilharzia brevis*
Hierarchy: Trichobilharzia: Schistosomatidae: Schistosomatoidea: Strigeidida: Digenea: Trematoda: Platyhelminthes
Trichoderma
Species: *Trichoderma viride, Trichoderma pseudokoningii*
Hierarchy: Trichoderma: Hypocreaceae: Hypocreales: Hypocreomycetidae: Sordariomycetes: Ascomycota
Trichomonas
Species: *Trichomonas vaginalis, Trichomonas tenax*
Hierarchy: Trichomonas: Trichomonadidae: Trichomonadida: Parabasalia: Metamonada
Trichophyton
Species: *Trichophyton violaceum, Trichophyton verrucosum, Trichophyton tonsurans, Trichophyton soudanense, Trichophyton simii, Trichophyton schoenleinii, Trichophyton rubrum, Trichophyton mentagrophytes, Trichophyton megninii, Trichophyton gourvilii, Trichophyton gallinae, Trichophyton equinum, Trichophyton concentricum, Trichophyton ajelloi*
Hierarchy: Trichophyton: Mitosporic arthrodermataceae: Arthrodermataceae: Onygenales: Eurotiomycetidae: Eurotiomycetes: Leotiomyceta: Pezizomycotina: Saccharomyceta: Ascomycota

Trichosporon

Species: *Trichosporon ovoides, Trichosporon mucoides, Trichosporon inkin, Trichosporon cutaneum, Trichosporon beigelii, Trichosporon asahii*

Hierarchy: Trichosporon: Mitosporic tremellales: Tremellales: Tremellomycetes: Agaricomycotina: Basidiomycota

Trichostrongylus

Species: *Trichostrongylus vitrinus, Trichostrongylus skrjabini, Trichostrongylus probolurus, Trichostrongylus orientalis, Trichostrongylus lerouxi, Trichostrongylus instabilis, Trichostrongylus colubriformis, Trichostrongylus capricola, Trichostrongylus calcaratus, Trichostrongylus brevis, Trichostrongylus axei, Trichostrongylus affinis*

Hierarchy: Trichostrongylus: Trichostrongylinae: Trichostrongylidae: Trichostrongyloidea: Strongylida: Rhabditida: Nematoda

Trichuris

Species: *Trichuris vulpis, Trichuris trichiura, Trichuris suis*

Hierarchy: Trichuris: Trichuridae: Trichocephalida: Enoplia: Enoplea: Nematoda

Tritirachium

Species: *Tritirachium oryzae*

Hierarchy: Tritirachium: Mitosporic ascomycota: Ascomycota

Tropheryma

Species: *Tropheryma whippelii*

Hierarchy: Tropheryma: Cellulomonadaceae: Actinomycetales: Actinobacteridae: Actinobacteria

Trypanosoma

Species: *Trypanosoma cruzi, Trypanosoma brucei*

Hierarchy: Trypanosoma: Trypanosomatidae: Kinetoplastida: Euglenozoa

Tsukamurella (pneumonia in immune-compromised patients)

Species: *Tsukamurella tyrosinosolvens, Tsukamurella pulmonis, Tsukamurella paurometabola, Tsukamurella inchonensis*

Hierarchy: Tsukamurella: Tsukamurellaceae: Corynebacterineae: Actinomycetales: Actinobacteridae: Actinobacteria (class): Actinobacteria

Tubercularia

Species: *Tubercularia vulgaris*

Hierarchy: Tubercularia: Mitosporic hypocreales: Hypocreales: Hypocreomycetidae: Sordariomycetes: Sordariomyceta: Leotiomyceta: Pezizomycotina: Saccharomyceta: Ascomycota

Ulocladium

Species: *Ulocladium chartarum*

Hierarchy: Ulocladium: Mitosporic pleosporaceae: Pleosporaceae: Pleosporineae: Pleosporales: Pleosporomycetidae: Dothideomycetes: Dothideomyceta: Leotiomyceta: Pezizomycotina: Saccharomyceta: Ascomycota

Uncinaria

Species: *Uncinaria stenocephala*

Hierarchy: Uncinaria: Ancylostomatinae: Ancylostomatidae: Ancylostomatoidea: Strongylida: Rhabditida: Nematoda

Ureaplasma

Species: *Ureaplasma urealyticum*

Hierarchy: Ureaplasma: Mycoplasmataceae: Mycoplasmatales: Mollicutes: Tenericutes

Veillonella

Species: *Veillonella parvula, Veillonella dispar, Veillonella atypica*

Hierarchy: Veillonella: Veillonellaceae: Selenomonadales: Negativicutes: Firmicutes

Veronaea

Species: *Veronaea botryosa*

Hierarchy: Veronaea: Mitosporic herpotrichiellaceae: Herpotrichiellaceae: Chaetothyriales: Chaetothyriomycetidae: Eurotiomycetes: Leotiomyceta: Pezizomycotina: Saccharomyceta: Ascomycota

Vibrio

Species: *Vibrio vulnificus, Vibrio parahaemolyticus, Vibrio mimicus, Vibrio hollisae, Vibrio furnissii, Vibrio fluvialis, Vibrio cincinnatiensis, Vibrio cholerae, Vibrio alginolyticus*

Hierarchy: Vibrio: Vibrionaceae: Vibrionales: Gamma Proteobacteria: Proteobacteria

Vittaforma

Species: *Vittaforma corneae*

Hierarchy: Vittaforma: Nosematidae: Apansporoblastina: Microsporidia

Volutella (endophthalmitis)

Species: *Volutella cinerescens* (formerly *Psilonia cinerescens*)

Hierarchy: Psilonia: Nectriaceae: Hypocreales: Hypocreomycetidae: Ascomycetes: Pezizomycotina: Ascomycota: Dikarya: Fungi

Watsonius

Species: *Watsonius watsoni*

Hierarchy: Watsonius: Paramphistomatidae: Paramphistomata: Echinostomida: Digenea: Trematoda: Platyhelminthes

Wolinella

Species: *Wolinella succinogenes*

Hierarchy: Wolinella: Helicobacteraceae: Campylobacterales: Epsilon Proteobacteria: Delta/epsilon subdivisions: Proteobacteria

Wuchereria

Species: *Wuchereria lewisi, Wuchereria bancrofti*

Hierarchy: Wuchereria: Onchocercidae: Filarioidea: Spirurina: Spirurida: Secernentea: Nematoda

Yersinia

Species: *Yersinia ruckeri, Yersinia rohdei, Yersinia pseudotuberculosis, Yersinia pestis, Yersinia mollaretii, Yersinia kristensenii, Yersinia intermedia, Yersinia frederiksenii, Yersinia enterocolitica, Yersinia bercovieri*

Hierarchy: Yersinia: Enterobacteriaceae: Enterobacteriales: Gamma Proteobacteria: Proteobacteria

References

1. Berman JJ. *Ruby programming for medicine and biology*. Sudbury: Jones and Bartlett; 2008.
2. *U.S. and World population clocks*. U.S. Census Bureau. <http://www.census.gov/main/www/popclock.html>; 2011 [accessed 20.7.11].
3. *The world factbook*. Washington, DC: Central Intelligence Agency; 2009.
4. *The state of world health*. In: Chapter 1, World Health Report 1996. World Health Organization; 1996.
5. *Weekly epidemiological record*. World Health Organization. 2007;**32**:285–96.
6. Muhlestein JB, Anderson JL. Chronic infection and coronary artery disease. *Cardiol Clin* 2003;**21**:333–62.
7. zur Hausen H. *Infections causing human cancer*. Hoboken: John Wiley and Sons; 2006.
8. Hatzakis A, Wait S, Bruix J, Buti M, Carballo M, Cavaleri M, et al. The state of hepatitis B and C in Europe: report from the hepatitis B and C summit conference. *J Viral Hepat* 2011;**18**(Suppl. 1):1–16.
9. Wales JH, Sinnhuber RO, Hendricks JD, Nixon JE, Eisele TA. Aflatoxin B1 induction of hepatocellular carcinoma in the embryos of rainbow trout (*Salmo gairdneri*). *J Natl Cancer Inst* 1978;**60**:1133–9.
10. Baron EJ, Allen SD. Should clinical laboratories adopt new taxonomic changes? If so, when? *Clin Infect Dis* 1993;**16**(Suppl. 4):S449–50.
11. Pearson K. *The grammar of science*. London: Adam and Black; 1900.
12. Scamardella JM. Not plants or animals: a brief history of the origin of Kingdoms Protozoa, Protista and Protoctista. *Internatl Microbiol* 1999;**2**:207–16.
13. Simpson GG. *Principles of animal taxonomy*. New York: Columbia University Press; 1961.
14. Mayr E. Two empires or three? *PNAS* 1998;**95**:9720–3.
15. Lane N. *Life ascending: The ten great inventions of evolution*. London: Profile Books; 2009.
16. DeQueiroz K. Ernst Mayr and the modern concept of species. *PNAS* 2005;**102**(Suppl. 1):6600–7.
17. Woese CR, Fox GE. Phylogenetic structure of the prokaryotic domain: the primary kingdoms. *PNAS* 1977;**74**:5088–90.
18. Koonin EV, Galperin MY. *Sequence, evolution, function: computational approaches in comparative genomics*. Boston: Kluwer Academic; 2003.
19. Woese CR. Default taxonomy: Ernst Mayr's view of the microbial world. *PNAS* 1998;**95**:11043–6.
20. Zuckerkandl E, Pauling L. Molecules as documents of evolutionary history. *J Theor Biol* 1965;**8**:357–66.
21. Woese CR. Bacterial evolution. *Microbiol Rev* 1987;**51**:221–71.
22. Beiko RG. Telling the whole story in a 10,000-genome world. *Biol Direct* 2011;**6**:34.
23. Philippe H, Brinkmann H, Lavrov DV, Littlewood DT, Manuel M, Worheide G, et al. Resolving difficult phylogenetic questions: why more sequences are not enough. *PLoS Biol* 2011;**9**:e1000602.
24. Bergsten J. A review of long-branch attraction. *Cladistics* 2005;**21**:163–93.

25. Conrad ME. *Ehrlichia canis*: a tick-borne rickettsial-like infection in humans living in the southeastern United States. *Am J Med Sci* 1989;**297**:35–7.
26. Malik A, Jameel MN, Sohail S, Mir S. Human granulocytic anaplasmosis affecting the myocardium. *J Gen Intern Med* 2005;**20**:958.
27. Slatko BE, Taylor MJ, Foster JM. The Wolbachia endosymbiont as an anti-filarial nematode target. *Symbiosis* 2010;**51**:55–65.
28. Resnikoff S, Pascolini D, Etyaale D, Kocur I, Pararajasegaram R, Pokharel GP, et al. Global data on visual impairment in the year 2002. *Bull World Health Organ* 2004;**82**:844–51.
29. Cowan G. Rickettsial diseases: the typhus group of fevers: a review. *Postgrad Med J* 2000;**76**:269–72.
30. Kannangara S, DeSimone JA, Pomerantz RJ. Attenuation of HIV-1 infection by other microbial agents. *J Infect Dis* 2005;**192**:1003–9.
31. Ahmed AA, Pineda R. *Alcaligenes xylosoxidans* contact lens-related keratitis – a case report and literature review. *Eye Contact Lens* 2011;**37**:386–9.
32. *Global prevalence and incidence of selected curable sexually transmitted infections: overview and estimates.* Geneva: World Health Organization; 2001.
33. Williams KP, Gillespie JJ, Sobral BW, Nordberg EK, Snyder EE, Shallom JM, et al. Phylogeny of gammaproteobacteria. *J Bacteriol* 2010;**192**:2305–14.
34. Vandepitte J, Lemmens P, DeSwert L. Human Edwardsiellosis traced to ornamental fish. *J Clin Microbiol* 1983;**17**:165–7.
35. Bagley ST. Habitat association of Klebsiella species. *Infect Control* 1985;**6**:52–8.
36. Bottone EJ. *Yersinia enterocolitica*: the charisma continues. *Clin Microbiol Rev* 1997;**10**:257–76.
37. Shigematsu M, Kaufmann ME, Charlett A, Niho Y, Pitt TL. An epidemiological study of *Plesiomonas shigelloides* diarrhoea among Japanese travellers. *Epidemiol Infect* 2000;**125**:523–30.
38. Dowling JN, Saha AK, Glew RH. Virulence factors of the family Legionellaceae. *Microbiol Rev* 1992;**56**:32–60.
39. Fry NK, Warwick S, Saunders NA, Embley TM. The use of 16s ribosomal RNA analyses to investigate the phylogeny of the family Legionellaceae. *J Gen Microbiol* 1991;**137**:1215–22.
40. Murphy TF, Parameswaran GI. *Moraxella catarrhalis*, a human respiratory tract pathogen. *Clin Infect Dis* 2009;**49**:124–31.
41. Schoenborn L, Abdollahi H, Tee W, Dyall-Smith M, Janssen PH. A member of the delta subgroup of Proteobacteria from a pyogenic liver abscess is a typical sulfate reducer of the genus Desulfovibrio. *J Clin Microbiol* 2001;**39**:787–90.
42. Warren JR, Marshall BJ. Unidentified curved bacilli on gastric epithelium in active chronic gastritis. *Lancet* 1983;**1**:1273–5.
43. Kidd M, Modlin IM. A century of *Helicobacter pylori*: paradigms lost-paradigms regained. *Digestion* 1998;**59**:1–15.
44. Brown LM. *Helicobacter pylori*: epidemiology and routes of transmission. *Epidemiol Rev* 2000;**22**:28–97.
45. Kiehlbauch JA, Tauxe RV, Baker CN, Wachsmuth IK. *Helicobacter cinaedi*-associated bacteremia and cellulitis in immunocompromised patients. *Ann Intern Med* 1994;**121**:90–3.
46. Rivas D. *Human parasitology, with notes on bacteriology, mycology, laboratory diagnosis, hematology and serology*. Philadelphia: W.B. Saunders; 1920.

47. Brooke CJ, Margawani KR, Pearson AK, Riley TV, Robertson ID, Hampson DJ. Evaluation of blood culture systems for detection of the intestinal spirochaete Brachyspira (Serpulina) pilosicoli in human blood. *J Med Microbiol* 2000;**49**:1031–6.
48. Stamm LV. Global challenge of antibiotic-resistant *Treponema pallidum. Antimicrob Agents Chemother* 2010;**54**:583–9.
49. Khatchadourian K, Ovetchkine P, Minodier P, Lamarre V, Lebel MH, Tapiero B. The rise of the rats: a growing paediatric issue. *Paediatr Child Health* 2010;**15**:131–4.
50. Margulis L, Schwartz KV. *Five Kingdoms: an illustrated guide to the phyla of life on earth. 3rd ed.* New York: W.H. Freeman; 1998.
51. Anelet Jacobs A, Chenia HY. Biofilm formation and adherence characteristics of an *Elizabethkingia meningoseptica* isolate from *Oreochromis mossambicus. Ann Clin Microbiol Antimicrob* 2011;**10**:16.
52. Aliyu SH, Marriott RK, Curran MD, Parmar S, Bentley N, Brown NM, et al. Real-time PCR investigation into the importance of *Fusobacterium necrophorum* as a cause of acute pharyngitis in general practice. *J Med Microbiol* 2004;**53**(Pt 10):1029–35.
53. Cano RJ, Borucki MK. Revival and identification of bacterial spores in 25- to 40-million-year-old Dominican amber. *Science* 1995;**268**:1060–4.
54. Guillet C, Join-Lambert O, Le Monnier A, Leclercq A, Mechaï F, Mamzer-Bruneel MF, et al. Human listeriosis caused by *Listeria ivanovii. Emerg Infect Dis* 2010;**16**:136–8.
55. LeVay S. *When science goes wrong. Twelve tales from the dark side of discovery.* New York: Plume; 2008. p. 160–80.
56. Facklam R. What happened to the Streptococci: overview of taxonomic and nomenclature changes? *Clin Microbiol Rev* 2002;**15**:613–30.
57. Tauch A, Kaiser O, Hain T, Goesmann A, Weisshaar B, Albersmeier A, et al. Complete genome sequence and analysis of the multiresistant nosocomial pathogen *Corynebacterium jeikeium* K411, a lipid-requiring bacterium of the human skin flora. *J Bacteriol* 2005;**187**:4671–82.
58. Gordon MA. The genus dermatophilus. *J Bacteriol* 1964;**88**:509–22.
59. Gillum RL, Qadri SM, Al-Ahdal MN, Connor DH, Strano AJ. Pitted keratolysis: a manifestation of human dermatophilosis. *Dermatologica* 1988;**177**:305–8.
60. Global leprosy situation. *World Health Organ Wkly Epidemiol Rec* 2006;**81**:309–16.
61. Relman DA, Schmidt TM, MacDermott RP, Falkow S. Identification of the uncultured bacillus of Whipple's disease. *N Engl J Med* 1992;**327**:293–301.
62. McNeil MM, Brown JM, Scalise G, Piersimoni C. Nonmycetomic *Actinomadura madurae* infection in a patient with AIDS. *J Clin Microbiol* 1992;**30**:1008–10.
63. Brocks JJ, Logan GA, Buick R, Summons RE. Archaean molecular fossils and the early rise of eukaryotes. *Science* 1999;**285**:1033–6.
64. Stechmann A, Hamblin K, Perez-Brocal V, Gaston D, Richmond GS, van der Giezen M, et al. Organelles in Blastocystis that blur the distinction between mitochondria and hydrogenosomes. *Curr Biol* 2008;**18**:580–5.
65. Tovar J, Leon-Avila G, Sanchez LB, Sutak R, Tachezy J, van der Giezen M, et al. Mitochondrial remnant organelles of Giardia function in iron-sulphur protein maturation. *Nature* 2003;**426**:172–6.
66. Tovar J, Fischer A, Clark CG. The mitosome, a novel organelle related to mitochondria in the amitochondrial parasite Entamoeba histolytica. *Mol Microbiol* 1999;**32**:1013–21.
67. Burri L, Williams B, Bursac D, Lithgow T, Keeling P. Microsporidian mitosomes retain elements of the general mitochondrial targeting system. *PNAS* 2006;**103**:15916–20.
68. Margulis L. Undulipodia, flagella and cilia. *Biosystems* 1980;**12**:105–8.

69. Cavalier-Smith T. The phagotrophic origin of eukaryotes and phylogenetic classification of Protozoa. *Int J Syst Evol Microbiol* 2002;**52**(Pt 2):297–354.
70. Soper D. Trichomoniasis: under control or undercontrolled? *Am J Obstet Gynecol* 2004; **190**:281–90.
71. Hobbs MM, Kazembe P, Reed AW, Miller WC, Nkata E, Zimba D, et al. *Trichomonas vaginalis* as a cause of urethritis in Malawian men. *Sex Transm Dis* 1999;**26**:381–7.
72. Stark D, Beebe N, Marriott D, Ellis J, Harkness J. Prospective study of the prevalence, genotyping, and clinical relevance of Dientamoeba fragilis infections in an Australian population. *J Clin Microbiol Infect Dis* 2005;**43**:2718–23.
73. Baldauf SL. An overview of the phylogeny and diversity of eukaryotes. *J Syst Evol* 2008; **46**:263–73.
74. Rassi A Jr, Rassi A, Marin-Neto JA. Chagas disease. *Lancet* 2010;**375**:1388–402.
75. Hampl V, Hug L, Leigh JW, Dacks JB, Lang BF, Simpson AGB, et al. Phylogenomic analyses support the monophyly of Excavata and resolve relationships among eukaryotic supergroups. *PNAS* 2009;**106**:3859–64.
76. Lemon SM, Sparling PF, Hamburg MA, Relman DA, Choffnes ER, Mack A. *Vector-borne diseases: understanding the environmental, human health, and ecological connections, workshop summary.* Institute of Medicine (US) Forum on Microbial Threats. Washington (DC): National Academies Press; 2008.
77. Cogswell FB. The hypnozoite and relapse in primate malaria. *Clin Microbiol Rev* 1992; **5**:26–35.
78. Joseph JT, Roy SS, Shams N, Visintainer P, Nadelman RB, Hosur S, et al. Babesiosis in Lower Hudson Valley, New York, USA. *Emerg Infect Dis* 2011;**17**:843–7.
79. Wong KT, Pathmanathan R. High prevalence of human skeletal muscle sarcocystosis in south-east Asia. *Trans Royal Soc Trop Med Hyg* 1992;**86**:631–2.
80. Stensvold CR, Suresh GK, Tan KS, Thompson RC, Traub RJ, Viscogliosi E, et al. Terminology for Blastocystis subtypes: a consensus. *Trends Parasitol* 2007;**23**:93–6.
81. Amin OM. Seasonal prevalence of intestinal parasites in the United States during 2000. *Am J Trop Med Hyg* 2002;**66**:799–803.
82. Lahr DJG, Grant J, Nguyen T, Lin JH, Katz LA. Comprehensive phylogenetic reconstruction of amoebozoa based on concatenated analyses of SSU-rDNA and actin genes. *PLoS ONE* 2011;**6**: e22780.
83. da Rocha-Azevedo B, Tanowitz HB, Marciano-Cabral F. Diagnosis of infections caused by pathogenic free-living amoebae. *Interdiscip Perspect Infect Dis* 2009;**2009**:251406.
84. Jian B, Kolansky AS, Baloach ZW, Gupta PK. *Entamoeba gingivalis* pulmonary abscess – diagnosed by fine needle aspiration. *Cytojournal* 2008;**5**:12.
85. Fredricks DN, Jolley JA, Lepp PW, Kosek JC, Relman DA. *Rhinosporidium seeberi*: a human pathogen from a novel group of aquatic protistan parasites. *Emerg Infect Dis* 2000;**6**:273–82.
86. Nowack EC, Melkonian M, Glockner G. Chromatophore genome sequence of Paulinella sheds light on acquisition of photosynthesis by eukaryotes. *Curr Biol* 2008;**18**:410–8.
87. Lass-Florl C, Mayr A. Human protothecosis. *Clin Microbiol Rev* 2007;**20**:230–42.
88. Leimann BC, Monteiro PC, Lazera M, Candanoza ER, Wanke B. Protothecosis. *Med Mycol* 2004;**42**:95–106.
89. Tyler S. Epithelium – the primary building block for metazoan complexity. *Integr Comp Biol* 2003;**43**:55–63.
90. Crompton DW. How much human helminthiasis is there in the world? *J Parasitol* 1999;**85**:397–403.

91. Foster J, Ganatra M, Kamal I, Ware J, Makarova K, Ivanova N, et al. The Wolbachia genome of *Brugia malayi*: endosymbiont evolution within a human pathogenic nematode. *PLoS Biol* 2005;**3**:e121.
92. Orihel TC, Eberhard ML. Zoonotic filariasis. *Clin Microbiol Rev* 1998;**11**:366–81.
93. Marcos LA, Terashima A, Dupont HL, Gotuzzo E. Strongyloides hyperinfection syndrome: an emerging global infectious disease. *Trans R Soc Trop Med Hyg* 2008;**102**:314–8.
94. Palmieri JR, Ratiwayanto S, Masbar S, Tirtokusumo S, Rusch J, Marwoto HA. Evidence of possible natural infections of man with *Brugia pahangi* in South Kalimantan Borneo Indonesia. *Trop Geogr Med* 1985;**37**:239–344.
95. Pozio E. New patterns of Trichinella infection. *Vet Parasitol* 2001;**98**:133–48.
96. Bruschi F, Murrell KD. New aspects of human trichinellosis: the impact of new Trichinella species. *Postgrad Med J* 2002;**78**:15–22.
97. Salehabadi A, Mowlavi G, Sadjjadi SM. Human infection with *Moniliformis moniliformis* in Iran: another case report after three decades. *Vector-borne and Zoonotic Dis* 2008; **8**:101–3.
98. Gaunt MW, Miles MA. An insect molecular clock dates the origin of the insects and accords with palaeontological and biogeographic landmarks. *Mol Biol Evol* 2002;**19**:748–61.
99. Hopkin SP. *The biology of the Collembola (springtails): The most abundant insects in the world.* London: Natural History Museum; 2007.
100. Kenney M, Eveland LK, Yermakov V, Kassouny DY. Two cases of enteric myiasis in man. Pseudomyiasis and true intestinal myiasis. *Am J Clin Pathol* 1976;**66**:786–91.
101. Tappe D, Buttner DW. Diagnosis of human visceral pentastomiasis. *PLoS Negl Trop Dis* 2009;**3**:e320.
102. Guarro J, Gene J, Stchigel AM. Developments in fungal taxonomy. *Clin Microbiol Rev* 1999;**12**:454–500.
103. Pounder JI, Simmon KE, Barton CA, Hohmann SL, Brandt ME, Petti CA. Discovering potential pathogens among fungi identified as nonsporulating molds. *J Clin Microbiol* 2007;**45**:568–71.
104. Prabhu RM, Patel R. Mucormycosis and entomophthoramycosis: a review of the clinical manifestations, diagnosis and treatment. *Clin Microbiol Infect* 2004;**10**(Suppl. 1):31–47.
105. Inamadar AC, Palit A. The genus Malassezia and human disease. *Indian J Dermatol Venereol Leprol* 2003;**69**:265–70.
106. Gueho E, Leclerc MC, de Hoog GS, Dupont B. Molecular taxonomy and epidemiology of Blastomyces and Histoplasma species. *Mycoses* 1997;**40**:69–81.
107. Walker J, Conner G, Ho J, Hunt C, Pickering L. Giemsa staining for cysts and trophozoites of *Pneumocystis carinii*. *J Clin Pathol* 1989;**42**:432–44.
108. Peraica M, Radic B, Lucic A, Pavlovic M. Toxic effects of mycotoxins in humans. *Bull World Health Organ* 1999;**77**:754–66.
109. Peres LC, Figueiredo F, Peinado M, Soares FA. Fulminant disseminated pulmonary adiaspiromycosis in humans. *Am J Trop Med Hyg* 1992;**46**:146–50.
110. Kobayashi M, Ishida E, Yasuda H, Yamamoto O, Tokura Y. Tinea profunda cysticum caused by *Trichophyton rubrum*. *J Am Acad Dermatol* 2006;**54**(2 Suppl.):S11–13.
111. Keeling PJ, Luker MA, Palmer JD. Evidence from beta-tubulin phylogeny that microsporidia evolved from within the fungi. *Mol Biol Evol* 2000;**17**:23–31.
112. Sak B, Kvac M, Kucerova Z, Kvetonova D, Sakova K. Latent microsporidial infection in immunocompetent individuals: a longitudinal study. *PLoS Negl Trop Dis* 2011;**5**:e1162.

113. Emerman M, Malik HS. Paleovirology: modern consequences of ancient viruses. *PLoS Biology* 2010;**8**:e1000301.
114. Angly FE, Felts B, Breitbart M, Salamon P, Edwards RA, Carlson C, et al. The marine viromes of four oceanic regions. *PLoS Biol* 2006;**4**:e368.
115. Horie M, Honda T, Suzuki Y, Kobayashi Y, Daito T, Oshida T, et al. Endogenous non-retroviral RNA virus elements in mammalian genomes. *Nature* 2010;**463**:84–7.
116. Bai L, Baker DR, Rivers M. Experimental study of bubble growth in Stromboli basalt melts at 1 atm. *Earth Planet Sci Let* 2008;**267**:533–54.
117. Koga Y, Kyuragi T, Nishihara M, Sone. N. Archaeal and bacterial cells arise independently from noncellular precursors? A hypothesis stating that the advent of membrane phospholipid with enantiomericglycerophosphate backbones caused the separation of the two lines of descent. *J Mol Evol* 1998;**46**:54–63.
118. Siebert C. Unintelligent design. *Discover Magazine* March 15, 2006.
119. Forterre P. Three RNA cells for ribosomal lineages and three DNA viruses to replicate their genomes: a hypothesis for the origin of cellular domain. *PNAS* 2006;**106**:3669–74.
120. Mackenzie JS, Field HE, Guyatt KJ. Managing emerging diseases borne by fruit bats (flying foxes), with particular reference to henipaviruses and Australian bat lyssavirus. *J Appl Microbiol* 2003;**94**:59S–69S.
121. Jern P, Sperber GO, Blomberg J. Use of Endogenous Retroviral Sequences (ERVs) and structural markers for retroviral phylogenetic inference and taxonomy. *Retrovirology* 2005;**2**:50.
122. Hughes AL, Friedman R. Poxvirus genome evolution by gene gain and loss. *Mol Phylogenet Evol* 2005;**35**:186–95.
123. Mohammed MA, Galbraith SE, Radford AD, Dove W, Takasaki T, Kurane I, et al. Molecular phylogenetic and evolutionary analyses of Muar strain of Japanese encephalitis virus reveal it is the missing fifth genotype. *Infect Genet Evol* 2011;**11**:855–62.
124. Magdi M, Hussein MM, Mooij JM, Roujouleh HM. Regression of post-transplant Kaposi sarcoma after discontinuing cyclosporin and giving mycophenolate mofetil instead. *Nephrol Dialysis Transplantat* 2000;**15**:1103–4.
125. Hensley LE, Mulangu S, Asiedu C, Johnson J, Honko AN, Stanley D, et al. Demonstration of cross-protective vaccine immunity against an emerging pathogenic Ebolavirus species. *PLoS Pathogens* 2010;**6**:e1000904.
126. Smallpox demise linked to spread of HIV infection. *BBC News* May 17, 2010.
127. Knight JC, Novembre FJ, Brown DR, Goldsmith CS, Esposito JJ. Studies on Tanapox virus. *Virology* 1989;**172**:116–24.
128. Arslan D, Legendre M, Seltzer V, Abergel C, Claverie J. Distant Mimivirus relative with a larger genome highlights the fundamental features of Megaviridae. *PNAS* 2011; **108**:17486–91.
129. La Scola B, Marrie T, Auffray J, Raoult D. Mimivirus in pneumonia patients. *Emerg Infect Dis* 2005;**11**:449–52.
130. Spiesschaert B, McFadden G, Hermans K, Nauwynck H, Van de Walle GR. The current status and future directions of myxoma virus, a master in immune evasion. *Vet Res* 2011;**42**:76.
131. Al-Rousan HO, Meqdam MM, Alkhateeb A, Al-Shorman A, Qaisy LM, Al-Moqbel MS. Human bocavirus in Jordan: prevalence and clinical symptoms in hospitalised paediatric patients and molecular virus characterisation. *Singapore Med J* 2011;**52**:365–9.

132. Akiba J, Umemura T, Alter HJ, Kojiro M, Tabor E. SEN virus: epidemiology and characteristics of a transfusion-transmitted virus. *Transfusion* 2005;**45**:1084–8.
133. Haruna I, Nozu K, Ohtaka Y, Spiegelman S. An RNA "replicase" induced by and selective for a viral RNA: isolation and properties. *PNAS* 1963;**50**:905–11.
134. Tolle MA. Mosquito-borne diseases. *Curr Probl Pediatr Adolesc Health Care* 2009; **39**:97–140.
135. Marx A, Glass JD, Sutter RW. Differential diagnosis of acute flaccid paralysis and its role in poliomyelitis surveillance. *Epidemiol Rev* 2000;**22**:298–316.
136. Nguyen M, Haenni AL. Expression strategies of ambisense viruses. *Virus Res* 2003; **93**:141–50.
137. Delgado S, Erickson BR, Agudo R, Blair PJ, Vallejo E, Albarino CG, et al. Chapare Virus, a newly discovered arenavirus isolated from a fatal hemorrhagic fever case in Bolivia. *PLoS Pathog* 2008;**4**:e1000047.
138. Bowen MD, Peters CJ, Mills JN, Nichol ST. Oliveros virus: a novel arenavirus from argentina. *Virology* 1996;**217**:362–6.
139. Gunsar F. Delta hepatitis. *Expert Rev Anti Infect Ther* 2009;**7**:499–501.
140. Forterre P. The two ages of the RNA world, and the transition to the DNA world: a story of viruses and cells. *Biochimie* 2005;**87**:793–803.
141. Prusiner SB. Novel proteinaceous infectious particles cause scrapie. *Science* 1982;**216**: 136–44.
142. Michelitsch MD, Weissman JS. A census of glutamine/asparagine-rich regions: implications for their conserved function and the prediction of novel prions. *PNAS* 2000;**97**:11910–5.
143. Traa BS, Walker CL, Munos M, Black RE. Antibiotics for the treatment of dysentery in children. *Int J Epidemiol* 2010;**39**(Suppl. 1):70–4.
144. Ecker DJ, Sampath R, Willett P, Wyatt JR, Samant V, Massire C, et al. The microbial rosetta stone database: a compilation of global and emerging infectious microorganisms and bioterrorist threat agents. *BMC Microbiol* 2005;**5**:19.
145. Taylor LH, Latham SM, Woolhouse ME. Risk factors for human disease emergence. *Philos Trans R Soc Lond B Biol Sci* 2001;**356**:983–99.
146. Berman JJ. *Methods in medical informatics: fundamentals of healthcare programming in Perl, Python, and Ruby*. Boca Raton: Chapman and Hall; 2010.
147. Schmidt GD. Acanthocephalan infections of man, with two new records. *J Parasitol* 1971;**57**:582–4.
148. Chai JY, Shin EH, Lee SH, Rim HJ. Foodborne intestinal flukes in Southeast Asia. *Korean J Parasitol* 2009;**47**:S69–S102.
149. Tada I, Otsuji Y, Kamiya H, Mimori T, Sakaguchi Y, Makizumi S. The first case of a human infected with an Acanthocephalan parasite, *Bolbosoma* sp. *J Parasitol* 1983; **69**:205–8.
150. Jueco NL, Monzon RB. *Cathaemasia cabrerai* sp. N. (Trematoda: Cathaemasiidae) a new parasite of man in the Philippines. *Southeast Asian J Trop Med Public Health* 1984; **15**:427–9.
151. Hasegawa H, Takao Y, Nakao M, Fukuma T, Tsuruta O, Ide K. Is *Enterobius gregorii* Hugot, 1983 (Nematoda: Oxyuridae) a distinct species? *J Parasitol* 1998;**84**:131–4.
152. Shoop WL. Experimental human infection with *Fibricola cratera* (Trematoda: Neodiplostomidae). *Kisaengchunghak Chapchi* 1989;**27**:249–52.

153. Quaiser A, Ochsenreiter T, Lanz C, Schuster SC, Treusch AH, Eck J, et al. Acidobacteria form a coherent but highly diverse group within the bacterial domain: evidence from environmental genomics. *Mol Microbiol* 2003;**50**:563–75.
154. Guk S, Park J, Shin E, Kim J, Lin A, Chai J. Prevalence of Gymnophalloides seoi infection in coastal villages of Haenam-gun and Yeongam-gun, Republic of Korea. *Korean J Parasitol* 2006;**44**:1–5.
155. Gardiner CH, Koh DS, Cardella TA. Micronema in man: third fatal infection. *Am J Trop Med Hyg* 1981;**30**:586–9.
156. Kaewkes S, Elkins DB, Haswell-Elkins MR, Sithithaworn P. *Phaneropsolus spinicirrus* n. sp. (Digenea: Lecithodendriidae), a human parasite in Thailand. *J Parasitol* 1991;**77**: 514–6.
157. Finegold SM, Jousimies-Somer H. Recently described clinically important anaerobic bacteria: medical aspects. *Clin Infect Dis* 1997;**25**(Suppl. 2):S88–93.
158. Nishizawa T, Okamoto H, Konishi K, Yoshizawa H, Miyakawa Y, Mayumi M. A novel DNA virus (TTV) associated with elevated transaminase levels in posttransfusion hepatitis of unknown etiology. *Biochem Biophys Res Commun* 1997;**241**:92–7.
159. Angert ER, Clements KD, Pace NR. The largest bacterium. *Nature* 1993;**362**:239–41.
160. Eby J. The prevalence of antibiotic-resistant bacteria on mosquitoes collected from a recreactional park. *Proceedings of the National Conference on Undergraduate Research* April 15–17, 2010.
161. Fang J. Ecology: a world without mosquitoes. *Nature* 2010;**466**:432–44.

Glossary

Bat A bat is not a flying mouse and is not a member of Class Rodentia, flying or otherwise. Bats are mammals of Class Chiroptera. With forelimbs that have evolved into wings, they are the only mammals capable of sustained, self-propelled flight. Their relevance in this book stems from their status as viral vectors (see Chapter 38). At present, bat populations of many species are being decimated by *Geomyces destructans*, a fungus in Class Ascomycota, the cause of white-nose syndrome. Apparently, the fungal infection, which grows in cold conditions, awakens bats from their deep hibernation; starvation results.

Blood contamination When a blood donor is infected with a pathogenic organism, the disease can be passed to the recipient. Examples of organisms and diseases that can be spread through transfused blood or blood components include:

Human immunodeficiency virus
Human T-lymphotropic viruses type I and type II
Hepatitis A
Hepatitis B
Hepatitis C
Hepatitis E
Cytomegalovirus
Epstein–Barr virus
Human parvovirus B19
Human herpesvirus 6
Human herpesvirus 8
TT virus or transfusion transmitted virus or torque teno virus [158]
SEN virus [132]
CJD and vCJD
Syphilis
Malaria
Chagas disease
African trypanosomiasis
Toxoplasmosis
Leishmaniasis
Babesiosis
Rocky Mountain spotted fever
Ehrlichiosis

Cladistics The technique of producing a hierarchy of clades, wherein each clade is a monophyletic class. See Monophyletic class.

Class A defined group within a taxonomy. The most familiar classes in biological taxonomy are the classes that form the ranked hierarchy of living organisms: Kingdom,

Phylum, Class, Order, Family, Genus, and Species. It is somewhat confusing that one of the classes of organisms is "Class," and another of the classes is named "Order." This means that when the terms "Class" or "Order" appear in a sentence, the reader must somehow distinguish between the general term and the specific term. In this book, classes are unranked. The word "class," lowercase, is used as a general term. The word "Class," uppercase, followed by an uppercase animal division (e.g. Class Animalia), represents a group within the taxonomy. In the biological hierarchy, each class has exactly one direct ancestor class (also called parent class or superclass), though an ancestor class can have more than one direct descendent class (also called child class, or subclass). See Taxonomic order.

Classification A hierarchical collection of classes in which each class inherits the properties of its parent class (superclass) and each species (class instance) occurs in exactly one class. A classification has five purposes: (1) To drive down the complexity of the knowledge domain; (2) As an informatics key that can be used to retrieve data and organize data pulled from diverse resources; (3) As a hypothesis-generating machine (i.e. always calling upon scientists to examine the relationships among classes of organisms); (4) To distill each class into a set of defining, essential properties that distinguish one class from all other classes; and (5) As a comprehensive collection of every member of a knowledge domain. Classifications should be distinguished from "Ontologies." An ontology is a system that relates groups of objects by rules. In an ontology, a class of objects may have more than one parent class (superclass) and an object may be a member of more than one class. A classification can be considered a special type of ontology wherein each class is limited to a single parent class and each object has membership in one and only one class. Classifications have simple hierarchical architectures; ontologies do not.

Commensal A symbiotic relationship in which one of the organisms benefits and the other is unaffected, under normal conditions. An opportunistic commensal is an organism that does not produce disease in its host, unless the host provides a physiologic opportunity for disease, such as malnutrition, advanced age, immunodeficiency, overgrowth of the organism (e.g. after antiobiotic usage), or some mechanical portal that introduces the organism to a part of the body that is particularly susceptible to the pathologic expression of the organism (such as an indwelling catheter, or an intravenous line). See Opportunistic infection.

Convergence When two species independently acquire an identical or similar trait through adaptation; not through inheritance from a shared ancestor. Examples are: the wing of a bat and the wing of a bird; the opposable thumb of opossums and of primates; the beak of a platypus and the beak of a bird.

Cyanobacteria The most influential organisms on earth, cyanobacteria, were the first and only organism to master the biochemical intricacies of oxygenic photosynthesis (more than 3 billion years ago). Photosynthesis involves a photochemical reaction, that uses carbon dioxide and water, and releases oxygen. All photosynthesizing life forms are either cyanobacteria, or they are eukaryotic cells (e.g. algae, plants) that have acquired chloroplasts (an organelle created in the distant past by endosymbiosis between a eukaryote and a cyanobacteria). Before the emergence of oxygen-producing cyanobacteria, the earth's atmosphere had very little oxygen.

Endotoxin These are toxins produced by bacteria that are part of the structure of the organism (i.e. not an excreted molecule as found in exotoxins). Most of the endotoxins are cell-wall lipopolysaccharides. These molecules can produce generalized inflammatory reactions when injected into humans (e.g. fever, drop in blood pressure, activation of the inflammation and blood cascades). Endotoxins are typically found on the outer membrane of the cell wall. Gram-negative bacteria have an outer membrane that impedes the entrance of the Gram stain. Gram-positive bacteria lack this outer membrane and provide easy access to the Gram stain. It is not surprising, therefore, that most endotoxins come from Gram-negative bacteria. A classic example of an endotoxin is found in meningococcemia, due to infection with *Neisseria meningitidis* (Beta Proteobacteria, Chapter 6).

Exotic diseases in the United States For many, the word "exotic" brings to mind all things strange and exciting. For clinical microbiologists, "exotic" refers to infectious diseases that have arrived via a distant geographic location. In the United States, a variety of exotic diseases have been introduced in the past few decades due, primarily, to two influences: global warming and global travel. The normal habitat of many disease vectors is a tropical climate. As the temperature of the planet rises, the geographic range for tropical vectors expands. As for travel, one of the most devastating disease vectors is the jet airplane. Jets transport infectious passengers and vectors. Can it be just a coincidence that the major corporations of the airline industry are called "carriers"? Here are a few of the exotic diseases that have been introduced or re-introduced into the USA, in the past few decades: West Nile fever, yellow fever, Mayaro fever, Dengue fever, Chikungunya, SARS, monkeypox, CJD/BSE, HIV/AIDS, Lassa fever, malaria, leishmaniasis, Chagas disease, cyclosporan gastroenteritis, cholera [76].

Facultative intracellular organism An organism that is capable of living, and reproducing inside or outside of cells. The term may apply to any organism, but most of the facultative intracellular organisms are bacteria. Example genera include: Brucella, Francisella, Histoplasma, Listeria, Legionella, Mycobacterium, Neisseria, and Yersinia. See Obligate intracellular organism.

HACEK A group of proteobacteria, found in otherwise healthy individuals, that are known to cause some cases of endocarditis, especially in children, and which do not grow well from cultured blood (due primarily to their slow growth rates). The term HACEK is created from the initials of the organisms of the group:

Haemophilus, particularly *Haemophilus parainfluenzae* Gamma Proteobacteria (Gamma Proteobacteria, Chapter 7)

Aggregatibacter, including *Aggregatibacter actinomycetemcomitans* and *Aggregatibacter aphrophilus* (Gamma Proteobacteria, Chapter 7)

Cardiobacterium hominis (Gamma Proteobacteria, Chapter 7)

Eikenella corrodens (Beta Proteobacteria, Chapter 6)

Kingella, particularly *Kingella kingae* (Beta Proteobacteria, Chapter 6)

Hemolytic syndromes Hemolytic syndromes are characterized by the destruction of red blood cells. Red cell destruction is caused by organisms that invade and rupture the cells (e.g. Plasmodium species, Babesia species), or that release chemicals that lyse red cells (e.g. hemolysins, listeriolysin O, rhamnolipid), or that induce an antigenic response against the patient's own red cells (post-streptococcal autoimmune hemolytic anemia). Hemolytic syndromes (red cell lysis) should be distinguished from hemorrhagic syndromes (vascular leakage), with which it can be confused clinically and phonetically.

Hemorrhagic fevers Hemorrhagic fevers are produced by an agent (often a virus) that elicits vasoactive mediators (e.g. kinins, histamine); thus increasing endothelial permeability (i.e. producing leaky vessels) and leading to hypovolemic shock (i.e. shock due to lack of blood in vessels). Typically, the vasoactive mediators are produced as a viral cytopathic effect of infected reticuloendothelial cells (e.g. macrophages). The pathogenesis varies somewhat from virus to virus and patient to patient. The presence or absence of liver involvement, CNS involvement, and DIC (disseminated intravascular coagulation) will alter the clinical course of disease. The hemorrhagic fever viruses come from Group IV (Chapter 42) or Group V (Chapter 43), the single-stranded RNA viruses. They have a non-human reservoir that is restricted to a specific geographic location, but can be spread by infected humans. The exception is Dengue hemorrhagic fever, which, aside from rare transfusion infections, seems to transmit via an insect vector. Rickettsia bacteria can produce non-viral hemorrhagic fevers. Scrub typhus, caused by *Orientia tsutsugamushi*, and transmitted by trombiculid mites, is another example of a non-viral fever. Here is a list of hemorrhagic fever viruses

- Arenaviridae (Group V, Chapter 43)
  - Lassa fever
  - Argentine hemorrhagic fevers (Junin virus)
  - Bolivian hemorrhagic fevers (Machupo virus)
  - Brazilian hemorrhagic fevers (Sabia virus)
  - Venezuelan hemorrhagic fevers (Guanarito virus)
- Bunyaviridae (Group V, Chapter 43)
  - Hantaviruses that cause hemorrhagic fever with renal syndrome (HFRS)
  - Nairovirus that causes Crimean-Congo hemorrhagic fever (CCHF)
  - Phlebovirus that causes Rift Valley fever (RVF)
- Filoviridae (Group V, Chapter 43)
  - Ebola hemorrhagic fever
  - Marburg hemorrhagic fever
- Flaviviridae (Group IV, Chapter 42)
  - Dengue fever (severe form)
  - Yellow fever
  - Tick-borne encephalitis group that cause Omsk hemorrhagic fever virus and Kyasanur Forest disease virus
- Arenaviridae (Group V, Chapter 43)
  - Lujo virus

Hepatitis viruses Several of the viruses that cause hepatitis are provided with names that are easy to remember but impossible to reconcile as a coherent biological class. These are hepatitis A, B, C, D, E, F, and G. Pathogenic viruses that attack any particular organ need not all belong to the same biological class, and the named hepatitis viruses are no exception, belonging to Groups IV, V, and VII. In addition, not all pathogenic viruses that infect the liver belong to the named hepatitis viruses. Yellow fever virus, which has killed millions of people throughout history, is a Group IV hepatitis virus (Chapter 42). Here is a list of the named hepatitis viruses:

Hepatitis A virus is a member of Class Picornaviridae (Group IV, Chapter 42).

Hepatitis B virus is a member of Class Hepadnaviridae (Group VII, Chapter 45).

Hepatitis C virus is a member of Class Flaviviridae (Group IV, Chapter 42), the same class that contains yellow fever virus, which also produces hepatitis.

Hepatitis D is a member of an unassigned class in Group V (Chapter 43).

Hepatitis E virus is a member of class Hepeviridae (Group IV, Chapter 42).

Hepatitis F virus is a hypothetical organism, supposedly responsible for some cases of hepatitis that cannot be diagnosed under any of the non-imaginary taxa.

Hepatitis G virus is now thought to be the same virus as GB virus C, a virus not known to produce any human disease.

Heterotrophic An organism is heterotrophic if it must acquire organic compounds from the environment as its energy source. All animals and all fungi (both of Class Opisthokonta) are heterotrophs. In contrast, members of Class Plantae, and other chloroplast-containing members of Class Eukaryota, are phototropic autotrophs; producing organic compounds from light, water, and carbon dioxide.

Infectious disease A disease caused by an organism that enters the human body. The term "infectious disease" is sometimes used in a way that excludes diseases caused by parasites. In this book, the parasitic diseases of humans are included among the infectious diseases.

Intermediate host Same as secondary host. A eukaryotic organism that contains a parasitic eukaryotic organism for a period of time during which the parasite matures in its life cycle, but in which maturation does not continue to the adult or sexual phase. Maturation to the adult or sexual phase occurs in the primary, or definitive host. A parasitic eukaryotic organism may have more than one intermediate host. The survival advantages offered to the parasite by the intermediate host stage may include the following: to provide conditions in which the particular stages of the parasite can develop, that are not available within the primary host; to disseminate the parasite (e.g. via water or air) to distant sites; to protect the immature forms from being eaten by the adult forms; to protect the parasite from harsh conditions that prevail in the primary host; to protect the parasite from external environmental conditions that prevail when the parasite leaves the primary host. See Primary host.

Kleptoplast Secondary chloroplast, wherein a host cell captures a chloroplast from another organism and uses the captured chloroplast as a temporary source of energy, until the chloroplast eventually ceases to function. Kleptoplasts typically capture chloroplasts throughout their lives, continuously replenishing exhausted organelles [73]. The only animal known to practice kleptoplasty is a type of sea slug (Class Mollusca), that ingests chloroplast-rich marine organisms.

Largest species For some time, the largest known bacterium was *Epulopiscium fishelsoni*, which grows to about 600 microns by 80 microns, much larger than the typical animal epithelial cell (about 35 microns in diameter) [159]. This record has since been exceeded by *Thiomargarita namibiensis*, a proteobacterium. This ocean-dwelling nonpathogen can reach a size of 0.75 millimeter, visible to the unaided eye. The largest single-celled organisms are deep-sea protists of Class Xenophyophorea, a subclass of

Class Rhizaria. These organisms can exceed four inches in length. Class Mimiviridae was thought to contain the largest viruses. Class Mimiviridae viruses can exceed 0.8 microns in length, thus exceeding the size of some bacteria (e.g. Mycoplasma species may be as small as 0.3 microns). The genome of species in Class Mimiviridae can exceed a million base pairs, encoding upwards of 1000 proteins. Class Megaviridae is a newly reported class of viruses, related to Class Mimiviridae, but bigger [128]. The organism with the largest genome is currently thought to be *Polychaos dubium* (Class Amoebozoa, Chapter 22), with a genome length of 670 billion base pairs; humans have a puny 3 billion base pair genome.

Long branch attraction When gene sequence data are analyzed, and two organisms share the same sequence in a stretch of DNA, it can be very tempting to infer that the two organisms belong to the same class (i.e. that they inherited the identical sequence from a common ancestor). This inference is not necessarily correct. Because DNA mutations arise stochastically over time, two species with different ancestors may achieve the same sequence in a chosen stretch of DNA. When mathematical phylogeneticists began modeling inferences for gene data sets, they assumed that most class assignment errors based on DNA sequence similarity would occur when the branches between sister taxa were long (i.e. when a long time elapsed between evolutionary divergences, allowing for many random substitutions in base pairs). They called this phenomenon, wherein non-sister taxa were assigned the same ancient ancestor class, "long branch attraction." In practice, errors of this type can occur whether the branches are long, or short, or in-between. Over the years, the accepted usage of the term "long branch attraction" has been extended to just about any error in phylogenetic grouping due to gene similarities acquired through any mechanism other than inheritance from a shared ancestor. This would include random mutation and adaptive convergence [24]. See Non-phylogenetic signal.

Metazoa Class Metazoa is equivalent to Class Animalia. It contains two subclasses: Class Parazoa and Class Eumetazoa. Class Eumetazoans contain all the animals that develop from a blastula. Class Parazoa contain two small subclasses: Class Porifera (the sponges) and Class Placozoa. Class Placozoa contains a single species, *Trichoplax adhaerens*. Sponges and *Trichoplax adhaerens* are exceedingly simple animals, consisting of a layer of jelly-like mesoderm sandwiched between simple epithelium.

Monophyletic class A class of organisms that includes a parent organism and all its descendants, while excluding any organisms that did not descend from the parent. If the subclasses of a parent class omit any of the descendants of the parent class, then the parent class is paraphyletic. If a subclass of a parent class includes organisms that did not descend from the parent, then the parent class is polyphyletic. A class can be paraphyletic and polyphyletic, if it excludes organisms that were descendants of the parent and if it includes organisms that did not descend from the parent. The goal of cladistics is to create a hierarchical classification that consists exclusively of monophyletic classes (i.e. no paraphyly, no polyphyly).

Mosquito Mosquitoes are members of Class Culicidae. Four genera of mosquitoes are vectors for human diseases: Aedes, Anopheles, Armigeres, and Culex. Among these genera, there are hundreds of individual species. Mosquitoes are vectors for biologically diverse organisms (animals, protists, and viruses). As yet, mosquitoes are not known to be vectors for bacterial diseases; but this biological oversight may soon be corrected. It has recently been shown that mosquitoes carry pathogenic bacteria, including antibiotic-resistant species [160]. Mosquitoes are reviled throughout the world, and have been likened to flying, infected hypodermic needles. The mosquito seems to serve no useful ecologic purpose (other than as a food source for bats, and other insectivores). It has been speculated that if every genera of mosquito were eliminated as a terrestrial species, there would be no significant negative ecologic repercussions [161]. The mosquito-borne virus diseases are discussed in Chapter 38. A few of the non-viral mosquito-transmitted diseases are listed here:

Malaria (Class Apicomplexa) is transmitted by species of Genus Anopheles.

Brugia malayii is transmitted by *Aedes polynesiensis* and other species.

Wuchereria bancrofti transmitted by *Armegeres subalbatus.*

Dermatobia hominis (human botfly) in which mosquitos carry the fly larvae, and the larvae follow the mosquito entry point into the host organism.

Negative classifier A negative classifier is a feature whose absence is used to place an organism into a taxonomic class; it is the riskiest way to assign classes. A species may lack a particular feature because none of its ancestors ever had the feature, as might be the case in a valid lineage of organisms. An example is the Collembola, popularly known as springtails, a ubiquitous member of Class Hexapoda, and easily found under just about any rock. These organisms look like fleas (same size, same shape) and were formerly included among the true fleas (Class Siphonaptera). Like fleas, springtails are wingless, and it was assumed that springtails, like fleas, lost their wings somewhere in evolution's murky past. However, true fleas lost their wings when they became parasitic. Springtails never had wings, an important taxonomic distinction separating springtails from fleas. Today, springtails (Collembola) are not classed with fleas or with any member of Class Insecta. They belong to Class Entognatha, a separate subclass of Class Hexapoda. Alternately, taxonomists may be deceived by a feature whose absence is falsely conceived to be a fundamental property of a class of organisms. For example, Class Fungi was believed to have a characteristic absence of a flagellum. Based on the absence of a flagellum, the fungi were excluded from Class Opisthokonta and were put in Class Plantae, which they superficially resembled. However, the chytrids, recently shown to be a primitive member of Class Fungi, have a flagellum. This places fungi among the true descendants of Class Opisthokonta.

Non-phylogenetic signal DNA sequences that cannot yield any useful conclusions related to the evolutionary position of an organism. Because DNA mutations arise stochastically over time (i.e. at random locations in the gene, and at random times), two organisms having different ancestors may achieve the same sequence in a chosen stretch of DNA. Long-branch attraction, mutational convergence, and adaptive convergence account for many of the errors that occur when non-phylogenetic signal is incorrectly assumed to have phylogenetic value [24].

Obligate intracellular organism An obligate intracellular organism can only reproduce within a host cell. Obligate intracellular organisms can include any type of organism, but the term aptly describes all viruses and all member of Class Chlamydia (Chapter 13). Examples of genera that contain obligate intracellular species include: Coxiella, Leishmania, Plasmodia, Rickettsia, Toxoplasma, Trypanosoma. See Facultative intracellular organism.

Opportunistic infection, Opportunistic organism Opportunistic infections are diseases that do not typically occur in healthy individuals, but which can occur in individuals who have a physiologic status favoring the growth of the organisms (e.g. diabetes, malnutrition). Sometimes, opportunistic infections occur in patients who are very old, or very young. Most often, opportunistic infections occur in immune-compromised patients. Specific diseases may increase susceptibility to specific types of organisms. For example, diabetics are more likely to contract systemic fungal diseases than are non-diabetic individuals. Some opportunistic infections arise from the population of organisms that live within most humans, without causing disease under normal circumstances (i.e. commensals). The concept of an opportunistic organism is, at best, a gray area of medicine, as virtually all of the organisms that arise in immune-compromised patients will, occasionally, cause disease in immune-competent patients (e.g. *Cryptococcus neoformans*). Moreover, the so-called primary infectious organisms, that produce disease in normal individuals, will tend to produce a more virulent version of the disease in immunosuppressed individuals (e.g. *Coccidioides immitis*). Examples of organisms that cause opportunistic infections are: *Acinetobacter baumanni* (Chapter 7), *Aspergillus* sp. (Chapter 36), *Candida* sp. (Chapter 36), *Clostridium difficile* (Chapter 12), *Cryptococcus neoformans* (Chapter 35), *Cryptosporidium parvum* (Chapter 19), cytomegalovirus (Chapter 39), herpes zoster (Chapter 39), *Histoplasma capsulatum* (Chapter 36), human herpesvirus 8 (Chapter 39), *Pneumocystis jirovecii* (Chapter 36), polyomavirus JC (Chapter 39), *Proteus* sp. (Chapter 7), *Pseudomonas aeruginosa* (Chapter 7), *Streptococcus pyogenes* (Chapter 12), *Toxoplasma gondii* (Chapter 19). See Commensals.

Parasite A parasite is an organism that lives and feeds in or on its host. In common usage, the term "parasite" is often reserved for animals that are parasitic in humans and other animals, and this has historically included the so-called one-celled animals. We now recognize that the so-called one-celled animals are distributed among divergent taxonomic classes. As we learn more and more about classes of organisms, the term "parasite" seems to have diminishing biologic specificity. In this book, the term "parasite" refers to any infectious organism. See Protozoa.

Primary host Also called final host or definitive host, the primary host is infected with the mature or reproductive stage of the parasite. In most cases, the mature stage of the parasite is the stage that produces eggs, larvae, or cysts. See Intermediate host.

Protozoa Microbiologic nomenclature has many terms that have persisted long after they have outlived their usefulness; "protozoa" is a perfect example. A commonly found definition for protozoa is "one-celled animal," but this is an oxymoron, as all animals are multicellular. Still, it is reasonable to assume that multi-animals must have evolved from unicellular organisms, and these one-celled prototypes animals could be called protozoans. If we adhered to this line of logic, only Class Choanozoa (Class 23) could lay claim as a "protozoan," as the choanozoans are one-celled members of Class Opisthokonta, to which Class Animalia belongs; most other so-called protozoans have no ancestral relationship to animals. A better way of thinking about "protozoan" is as a term that describes any one-celled heterotrophic eukaryotic organisms (i.e. members of Class Eukaryota that lack chloroplasts). With luck, the term "protozoa" will cease to appear in the scientific literature.

Sandfly Sandflies are small dipterans (flies), of several different genera, that live in sandy areas. The sandfly of genus Phlebotomus transmits Leishmania species (leishmaniasis, Chapter 17) and the phleboviruses that cause Pappataci fever (Chapter 43). The sandfly of genus Lutzomyia transmits *Bartonella bacilliformis* (bartonellosis, Chapter 5).

Secondary host Synonymous with intermediate host. See Intermediate host.

Serotype Subtypes of a species of bacteria or virus that differ in their surface antigens.

Serovar Same as serotype. See Serotype.

Taxonomic order In traditional taxonomy, the hierarchical lineage of organisms is divided into a descending list of named orders: Kingdom, Phylum (Division), Class, Order, Family, Genus, Species. In recent times, taxonomists have deemed it necessary to add many more formal categories (e.g. supraphylum, subphylum, suborder, infra-class, etc.). In some cases, where no named subdivision exists, a new class is created and given an "unranked" order. In many cases, the reason for these additional divisions relates to the need to impose monophyly on classes. A good example is the unranked class, Discoba. Molecular phylogenetic evidence has come to light suggesting that Class Jakobid is closely related to Class Percolozoa and Class Euglenozoa. Because these classes (Jakobid, Percolozoa, Euglenozoa) seem to have a common direct ancestor, they need to be assigned a common superclass. In this particular case, a superclass for these three sister classes was invented: Class Discoba. Because no named subdivision (i.e. rank) was available for Class Discoba, it is considered an "unranked" class. Aside from being an undue burden upon students who are trying to understand the practical aspects of biological taxonomy, it seems somewhat absurd to have a system larded with orders given an official label of "unranked." In this book, all classes are simply referred to as order "Class," followed by the name of the class. Each class has one named parent class. When you know the name of the parent for each class, you can determine the complete ancestral lineage for every class and species within the classification.

Taxonomy The science of classification, derived from the ancient Greek taxis, "arrangement," and nomia, "method." Naturalists use the word "taxonomy" to include the hierarchy of ancestral organisms and their descendants, and the names assigned to the classes and species of organisms. The product of the taxonomic effort is a classification, and the nomenclature for the species and classes is collectively called "the taxonomy." When used to describe the general field of classification, "taxonomy" is synonymous with "systematics."

Tick Ticks are members of Class Chelicerata (Chapter 29) and should not be confused with insects (Chapter 30). The hierarchy for the tick, Ixodes, is: Ixodes: Ixodinae: Ixodidae: Ixodoidea: Ixodida: Parasitiformes: Acari: Arachnida: Chelicerata: Arthropoda. Species *Ixodes scapularis* transmits *Babesia microti* (babesiosis), *Borrelia burgdorferi* (Lyme disease), and *Anaplasma phagocytophilum* (human granulocytic anaplasmosis) [78]. Tick-borne viruses include: Crimean-Congo hemorrhagic fever, tick-borne encephalitis, Powassan encephalitis, deer tick virus encephalitis, Omsk hemorrhagic fever, Kyasanur Forest disease (Alkhurma virus), Langat virus, and Colorado tick fever.

Vector An organism that moves a disease-causing organism from one host to another. Diseases spread by vectors include: malaria, plague, leishmaniasis, African trypanosomiasis, relapsing fever, yellow fever, dengue fever and dengue hemorrhagic fever, hantavirus disease, West Nile encephalitis, Japanese encephalitis, Rift Valley fever, Venezuelan equine encephalitis, and chikungunya. All arboviruses have arthropod vectors, and there are about 100 known arboviruses that cause human disease [76]. One vector can carry more than one type of infectious organism. For example, a single species of Anopheles mosquito can transmit *Dirofilaria immitis*, O'nyong'nyong fever virus, *Wuchereria bancrofti*, and *Brugia malayi*. Obversely, one disease organism can be spread by more than one vector. For example, orbiviruses are spread by mosquitoes, midges, gnats, sandflies, and ticks.

Virulence factor These are molecules that enhance the ability of an infectious organism to survive in its host. Many of the best-studied virulence factors are produced by bacteria, and because virulence factors always work to the detriment of the host, the terms "virulence factor" and "bacterial toxin" are mistakenly used interchangeably. You will find it useful to separate these two concepts because virulence factors and bacterial toxins have very different functions. Toxins work by damaging or killing host cells. Virulence factors work by making it easier for organisms to invade, grow, and persist within the host. Virulence is attained by helping the infectious organism obtain nutrition from host cells, by evading or suppressing the host immune response, by enhancing the ability of the organism to adhere to host cells, or by enabling the organisms to enter host cells or to invade through host tissues. See Toxin.

Zoonosis An infectious disease of humans that is acquired from a non-human animal reservoir. The method of infection (e.g. vector) does not determine whether a disease is considered to be zoonotic. Most fungal diseases are non-zoonotic, because fungi typically grow in soil or water, and are delivered to humans as airborne spores. For example, malaria, passed by a mosquito vector, is not a zoonosis, because the reservoir for organism that causes human malaria is usually another human. If there were no human carriers of malaria, the incidence of malaria would drop to insignificance. The same is true for schistosomiasis, river blindness, and elephantiasis. Though these diseases are transmitted by non-human vectors, their typical reservoir is human. Examples of zoonotic diseases are:

Anthrax
Babesiosis
Balantidiasis
Barmah Forest virus disease
Bartonellosis
Bilharzia
Bolivian hemorrhagic fever
Borna virus infection
Borreliosis (Lyme disease and others)
Bovine tuberculosis
Brucellosis
Campylobacteriosis
Cat scratch disease
Chagas disease
Cholera
Cowpox
Creutzfeldt–Jakob disease (vCJD)
Crimean–Congo hemorrhagic fever
Cryptosporidiosis
Cutaneous larva migrans
Dengue fever

Ebola fever
Echinococcosis
Erysipeloid
Escherichia coli O157:H7
Giardiasis
Glanders
H1N1 flu
Hantavirus infection
Helminthic infections
Hendra virus infection
Henipavirus infection
Korean hemorrhagic fever
Kyasanur forest disease
Lassa fever
Leishmaniasis
Leptospirosis
Listeriosis
Lyme disease
Lymphocytic choriomeningitis
Marburg fever
Mediterranean spotted fever
Monkey B virus infection
Nipah fever
Ocular larva migrans
Omsk hemorrhagic fever
Orf (animal disease)
Oropouche fever
Pasteurellosis
Plague
Psittacosis
Puumala virus infection
Q-fever
Rabies
Rift Valley fever
Ringworm from *Microsporum canis*
Rotavirus (when transmitted by dogs, cats and other animals)
Salmonellosis
SARS
Sodoku
Sparganosis
Streptococcus suis infection
Toxocariasis
Toxoplasmosis
Trichinosis

Tularemia
Typhus
Venezuelan hemorrhagic fever
Visceral larva migrans
West Nile fever
Western equine encephalitis virus
Yellow fever (sylvatic cycle)
Yersiniosis

Index

Note: Page numbers in **bold** indicate glossary entries.

D

E

F

G

U

V

W

Printed in the United States
By Bookmasters

Printed in the United States
By Bookmasters